CIRRHOSIS,
HYPERAMMONEMIA,
AND HEPATIC
ENCEPHALOPATHY

ADVANCES IN EXPERIMENTAL MEDICINE AND BIOLOGY

CIRRHOSIS, HYPERAMMONEMIA, AND HEPATIC ENCEPHALOPATHY

Edited by

Santiago Grisolía and Vicente Felipo

Instituto de Investigaciones Citologicas
Valencia, Spain

SPRINGER SCIENCE+BUSINESS MEDIA, LLC

Library of Congress Cataloging-in-Publication Data

Cirrhosis, Hyperammonemia, and Hepatic Encephalopathy / edited by
Santiago Grisolía and Vicente Felipo.
    p.   cm. -- (Advances in experimental medicine and biology ; v.
341)
    Includes bibliographical references and index.
    ISBN 978-1-4613-6058-2   ISBN 978-1-4615-2484-7 (eBook)
    DOI 10.1007/978-1-4615-2484-7
    1. Liver--Cirrhosis--Congresses. 2. Ammonia--Metabolic
detoxication--Congresses. 3. Urea--Metabolism--Congresses.
4. Hepatic encephalopathy--Congresses.   I. Grisolía, Santiago,
1923-   . II. Felipo, Vicente. III. Series.
RC848.C5C56 1993
616.3'62--dc20   93-40031
                                                              CIP

Proceedings of an International Summer Course on Cirrhosis, Hyperammonemia, and Hepatic
Encephalopathy, held August 10–14, 1992, in El Escorial, Spain

ISBN 978-1-4613-6058-2

©1993 Springer Science+Business Media New York
Softcover reprint of the hardcover 1st edition 1993
Originally published by Plenum Press in 1993

PREFACE

This volume contains the papers presented in the International Summer Course on "Cirrhosis, Hyperammonemia and Hepatic Encephalopathy," which was one of the prestigious Summer Course, of the Complutense University of Madrid held in El Escorial, Spain, during August 10-14, 1992.

Liver cirrhosis is one of the main causes of death in western countries. In addition there is a series of liver dysfunctions including fulminant hepatic failure, Reye's syndrome and congenital defects of urea cycle enzymes that could lead to hepatic encephalopathy, coma and death.

As a consequence of impaired liver function, the ability to detoxify ammonia by its incorporation into urea is diminshed, resulting in increased ammonia levels in blood and brain. Hyperammonemia is considered one of the main factors in the mediation of hepatic encephalopathy and the classical clinical treatments are directed towards reducing blood ammonia levels.

A part of the book is therefore devoted to the study of certain aspects of ammonia metabolism such as the regulation of the urea cycle, the main mechanism of ammonia detoxification in mammals, which is located mainly in the liver. The metabolism of ammonia in other tissues, including brain, is also presented, as well as the effects of hyperammonemia on brain metabolism and function and on brain microtubules. The control of cerebral protein breakdown is reviewed.

The classical and some recently proposed clinical treatments as well as nutritional considerations in the management of patients with liver failure are also discussed.

Altered neurotransmission is a key step in the pathogenesis of hepatic encephalopathy. Some recent results on the effects of hyperammonemia and hepatic encephalopathy on synaptic transmission and on GABAergic and glutamatergic neurotransmission are presented and their possible roles in the pathogenesis of hepatic encephalopathy are discussed.

Glutamatergic neurotransmission is altered in hyperammonemia and hepatic encephalopathy and it has been recently proposed that acute ammonia toxicity is mediated by glutamate, which, under certain conditions, can be a potent neurotoxic for neurons containing glutamate receptors. The last part of the book is devoted to the study of the mechanisms that control the release of glutamate, the properties of the NMDA receptors, which mediate glutamate toxicity, as well as the mechanism of the protective effect of gangliosides against glutamate toxicity.

In summary, the book presents an update of the knowledge on certain crucial aspects of the causes and mechanisms of hepatic encephalopathy.

We would like to express our gratitude to all the participants in the Course for their written contributions and for their enlightened and fruitful discussion and to Forpax for help in preparing the camera-ready version of the chapters. Also the Complutense University of Madrid for providing the financial support and the facilities for the Course and to its personnel at El Escorial for the excellent assistance and organization that allowed the fruitful development of the Course in a warm and pleasant atmosphere.

Santiago Grisolía
Vicente Felipo

CONTENTS

Control of Urea Synthesis and Ammonia Detoxification

Santiago Grisolía, María–Dolores Miñana, Eugenio Grau and Vicente Felipo

1. Introduction

Ammonia toxicity was first reported by Pavlov and coworkers a century ago (1). They excluded the liver from the circulation and found that when dogs treated in this way were fed meat, they developed hyperammonemia, which was associated with coma and led to the death of the animal.

Ammonia is produced in a great number of metabolic reactions, and is itself a precursor of a number of compounds. However, it is toxic when it reaches ≈ 4–5 times the normal levels. Ureotelic species get rid of ammonia mainly in the liver, via the urea cycle. The ammonia generated in most extrahepatic tissues is detoxified by incorporation into glutamine. This glutamine may be used in a number of biosynthetic reactions or released to the blood. Most of the ammonia incorporated into glutamine in these tissues however, is also finally eliminated as urea in the liver (2). Obviously, under normal conditions, the urea cycle plays a key role in the maintenance of whole body nitrogen homeostasis.

This equilibrium is seriously compromised in such liver diseases as fulminant hepatic failure and cirrhosis. There are also certain other conditions which are associated with hyperammonemia such as hereditary deficiencies of the urea cycle enzymes, Reye's syndrome, valproate therapy, organic aciduria, etc. (3, 4). The high ammonia levels reached under these conditions could lead to neurological complications, including coma and death. It is therefore important to know how nitrogen homeostasis is regulated under normal conditions as well as the metabolic adaptations occurring in hyperammonemic states.

2. Regulation of the Urea Cycle

As mentioned above, ammonia is detoxified mainly in the liver by incorporation into urea. This process is catalyzed by the so called urea cycle enzymes. The urea cycle was first proposed by Krebs and Henseleit in 1932 (5) and the first crucial finding leading to this proposal was the discovery of the exceptionally high rates of urea synthesis when both ornithine and ammonium ions were added to liver slices (6). The details of the enzymatic steps of the cycle were later clarified in the laboratories of Cohen and Ratner. There are five

Instituto de Investigaciones Citologicas, Amadeo de Saboya, 4. 46010 Valencia. Spain

Cirrhosis, Hyperammonemia, and Hepatic Encephalopathy,
Edited by S. Grisolia and V. Felipo, Plenum Press, New York, 1994

enzymes involved in the urea cycle; two of them are mitochondrial, the carbamylphosphate synthetase and the ornithine transcarbamylase while the other three (argininesuccinate synthetase, argininesuccinate lyase and arginase) are located in the cytosol. It was also shown that the first enzyme of the cycle (carbamylphosphate synthetase) requires the presence of an activator (7, 8). The physiological activator of carbamylphosphate synthetase (CPS) is acetylglutamate.

Two main mechanisms controlling the urea cycle in response to variations in dietary nitrogen intake have been described. There is a long–term adaptation in which the hepatic content of the five urea cycle enzymes responds to the protein intake and is reflected in the rate of urea excretion (9–11). There is also a short–term regulatory mechanism in which acetylglutamate, the physiological allosteric activator of CPS, plays a key role. Acetylglutamate levels in liver change rapidly in response to protein ingestion (12), arginine administration (13) or amino acid loads (14) and the variations in acetylglutamate levels correlate well with the ureogenic capacity of the animal (12).

3. Use of Ammonia Containing Diets to Study the Regulation of the Urea Cycle

Activities of the five urea cycle enzymes are depressed in cirrhotic human liver, as is the rate of urea synthesis (15–17). Average depression of activity per milligram of DNA was 54%, 37%, 50%, 64% and 69% for carbamylphosphate synthetase, ornithine transcarbamylase, argininosuccinate synthetase, argininosuccinate lyase, and arginase, respectively (15). It is therefore important to understand the regulation of the urea cycle (and its response to hyperammonemias induced by different means) in order to try to improve urea production in cirrhotic patients as well as in other hyperammonemic conditions.

It is well known that increasing dietary protein intake induces an elevation in all five urea cycle enzymes in the liver (9–11). However, this long–term adaptation is slow and it takes several days to reach the new equilibrium. It has been reported that ingestion of high protein diets raises the glucagon levels (18) and that glucagon stimulates the synthesis of urea cycle enzymes (19) by increasing their transcription rates (20). This suggests that the slow adaptative increase in urea cycle enzymes in liver in response to high nitrogen intake in the form of protein is mediated by hormonal changes.

We studied the effects on the urea cycle of increasing the intake of nitrogen in the form of ammonium ions or of protein. To do this we used a new animal model of hyperammonemia (21) consisting of feeding rats a standard diet (20% protein) supplemented with ammonium acetate (20%, w/w). We compared the effect of the ammonium diet with that of a diet containing the same amount of nitrogen in the form of protein (50% protein = high protein). We also tested the effect of a protein–free diet supplemented or not with 20% ammonium acetate (22).

The effects of these different diets on urea and orotic acid excretion are shown in Fig. 1. The urea excretion in animals fed the standard diet plus ammonium acetate reached 2.3–fold over that of controls. In addition, nearly all ammonium ingested was excreted as urea. We studied the mechanism controlling the urea cycle in response to ammonium ingestion. We found that rats fed the ammonium diet showed no increase in activity (Table 1) or amount (Fig. 2) of carbamylphosphate synthetase, while the high protein group did. These results indicate that the adaptative mechanisms are different when ammonia is derived from protein or provided as ammonium salts.

The activity of carbamylphosphate synthetase shown in Table 1 was determined under optimal conditions and can therefore be taken as an estimate of its maximum capacity for carbamyl–phosphate production. It is well known however, that the activity of CPS in vivo under standard dietary conditions is much lower than its maximum capacity and that it is

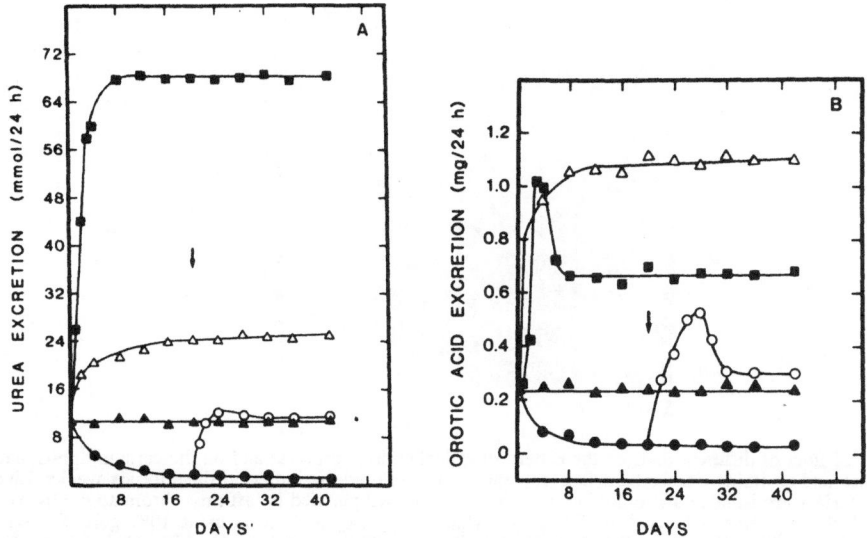

Figure 1. Effects of different diets on urea and orotic acid excretion. Rats were fed the following diets: (▲) standard; (△) standard plus ammonium; (•) protein–free; (◦) protein–free plus ammonium and (■) high protein (50%). The rats of the protein–free plus ammonium group were fed the protein–free diet for 20 days and then (indicated by arrow) transferred to the ammonium containing diet. Urine was collected each 24 hours and urea (A) and orotic acid (B) were determined. Each point is the mean of the values for four rats. From (22).

Table 1. Effect of several diets on the activity of carbamylphosphate synthetase I

DIET	Liver mass	Activity	
	g	U/mg prot.	U/liver
Standard	13.9 ± 0.9	4.9 ± 0.3	2020
Standard plus ammonium	13.6 ± 0.8	4.8 ± 0.2	1990
50% protein	16.0 ± 1.2	10.6 ± 1.0	4860
Protein–free	4.8 ± 0.5	2.2 ± 0.1	400
Protein–free plus ammonium	5.5 ± 0.9	2.2 ± 0.1	490

Rats were fed the indicated diets for 42 days, then killed and liver mitochondria were isolated. Activity/mass protein refer to mitochondrial protein. Values are the mean of six rats. From (22).

regulated essentially by the level of acetylglutamate, its physiological activator (14).

The level of acetylglutamate (AG) in liver mitochondria was therefore determined. As shown in Fig. 3, ingestion of ammonia increased the hepatic acetylglutamate by ≈ two–fold over the control values for standard diet and ≈ 8–fold for the protein–free diet. As shown in Fig. 1, urea excretion was also increased ≈ two–fold and 8–fold when standard or protein–free diets were supplemented with 20% ammonium acetate. This suggests that there is a good correlation between the amount of acetylglutamate and the synthesis of urea and that the amount of CPS does not play a key role in the control of ureagenesis. The amount of CPS/liver is reduced 5–fold in rats fed the protein–free diet (Table 1, Fig. 2). Under these conditions an 8–fold increase in the levels of AG leads to an 8–fold increase in urea synthesis, indicating that the amount of CPS is in a great excess over that necessary to meet the requirements for urea synthesis and that this is actually regulated by the levels of acetylglutamate.

We then studied how ammonia affects the hepatic level of acetylglutamate. The rise in AG levels could be due to increased synthesis, decreased breakdown or both. As shown

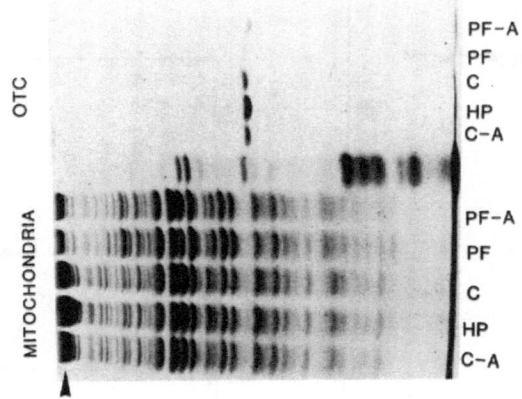

Figure 2. Effect of different diets on the pattern of mitochondrial proteins and on the carbamylphosphate synthetase and ornithine transcarbamylase contents. Rats were fed the indicated diets for six weeks. Liver mitochondria were isolated and ornithine transcarbamylase was purified by affinity chromatography with δ-PALO. Samples were subjected to SDS-polyacrylamide gel electrophoresis using 10% gels. The same amount of mitochondrial protein was applied to each lane. The diets used were as follows: C: standard diet; C-A standard plus ammonium; PF: protein-free; PF-A: protein-free plus ammonium; HP: high protein (50%). Arrow indicates the position of carbamylphosphate synthetase. From (32).

in Table 2, rats fed the ammonium diet did not have an increase in acetylglutamate activity but rather a slightly reduced activity (86% of control). Also the level of its activator, synthetasearginine, was slightly reduced. Moreover, as shown in Table 3, AG synthesis by mitochondria isolated from livers of rats on the ammonium diet was less than in controls. This indicate that synthesis of AG is not increased in rats fed the ammonium diet and suggests that the increased content of AG must be due to decreased breakdown.

The AG deacylating activity in liver is located in the cytosol (23) and degradation of

Table 2. Effect of ammonium ingestion on the activity of acetylglutamate synthase and on the liver content of its substrates and activators

DIET	Activity of acetylglutamate	Acetyl-CoA	Glutamate	Arginine
	cpm/mg prot.x h		nmol/g liver	
Standard	2040 ± 140	15 ± 1	1910 ± 180	204 ± 15
Standard plus ammonium	1760 ± 170	22 ± 2	1566 ± 160	151 ± 11

Rats were fed the standard or standard plus ammonium diet for six weeks. Mean of four duplicate determinations from four rats per group. From (22).

Table 3. Synthesis of acetylglutamate by mitochondria isolated from livers of rats on standard or standard plus ammonium diets

DIET	Acetylglutamate (pmol/mg protein)		
	Initial	Final	Synthesis
Standard	920 ± 70	1570 ± 120	650
Standard plus ammonium	1760 ± 110	2340 ± 150	580

Mitochondria (25 mg of protein) isolated from livers of rats on standard or standard plus ammonium diets were incubated for 60 min at 37°C with continuous shaking. Acetylglutamate was determined by activation of carbamylphosphate synthetase I and also by HPLC. From (22).

AG depends on the rate of efflux from the mitochondria. The effects of the ammonium diet on the rate of efflux of AG from the mitochondria are shown in Table 4. Under the experimental conditions used, mitochondria from control animals released ≈ 60% of the initial AG while those from the ammonium group released only 20% of the initial AG.

Table 4. Effect of ammonium ingestion on the efflux of acetylglutamate from mitochondria

Time of measurement	Acetylglutamate (pmol/mg protein)	
	Standard diet	Ammonium diet
Initial	850 ± 80	1720 ± 120
After 30 min with ammonium	2600 ± 130	3370 ± 220
Remaining in mitochondria		
after 2nd incubation	1020 ± 70	2810 ± 100
Final in incubation medium	1730 ± 120	690 ± 60

Mitochondria (25 mg protein) isolated from livers of rats on standard or standard plus ammonium diet were incubated at 37°C in a solution containing 4 mM ammonium acetate to increase the mitochondrial content of acetylglutamate. After 30 min, mitochondria were pelleted by centrifugation and resuspended in the same medium without ammonium acetate to follow the release of acetylglutamate. After 30 min mitochondria were pelleted and acetylglutamate was determined in both mitochondria and supernatant. Mean of three experiments. From (22).

These results indicate that ingestion of large amounts of ammonium (which makes the rats hyperammonemic) increases the hepatic acetylglutamate by reducing its efflux from the mitochondria and, therefore, its degradation. A similar mechanism appears to be involved in the increase in hepatic acetylglutamate induced by high protein diets (24).

In conclusion, when 20% ammonium acetate was added to a standard diet both the AG content and urea synthesis increased by ≈ two-fold; when ammonium acetate was added to a protein−free diet both the AG content and urea synthesis increased by ≈ 8−fold (Figs. 1 and 3); however, addition of ammonium did not affect the amount of CPS (Fig. 2, Table 1). These experiments therefore support the idea that urea synthesis is mainly controlled by the hepatic content of acetylglutamate.

4. Effects on the Urea Cycle of Protein Deprivation and Refeeding.

The above results were obtained using a model of hyperammonemia consisting of feeding rats an ammonia containing diet. Rats fed a protein−free diet also became hyperammonemic (Fig. 4) while the hepatic content of acetylglutamate and urea excretion were reduced by 84% (Fig. 3) and 89% (Fig. 1), respectively.

We studied the time−course of the changes in blood ammonia and urea levels, urea excretion, CPS activity and acetylglutamate content in liver following administration of the protein−free diet. After 20 days on this diet, some of the rats were refed the standard diet (20% protein) and the time−course of the recovery of the above urea cycle parameters was also determined (25).

As shown in Fig. 4A, blood ammonia decreased ≈ 30% the first day on the protein−free diet but increased thereafter for about two weeks, to reach 165% of control. Blood ammonia increased further the first day of refeeding the standard diet, decreased sharply during the following days and returned to control values after the third day. The effects of the protein−free diet and refeeding on the levels of urea in serum are shown in Fig. 4B. There was a rapid decrease during the initial 3 days on the protein−free diet followed by a stabilization. Urea in serum returned to control values after two days of refeeding the standard diet.

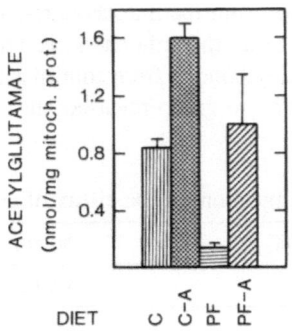

Figure 3. Acetylglutamate content in mitochondria from livers of rats on different diets. Rats were fed the indicated diets for six weeks. Mitochondria were isolated and acetylglutamate content was determined by activation of carbamylphosphate synthetase. Values are the mean of seven determinations. From (22).

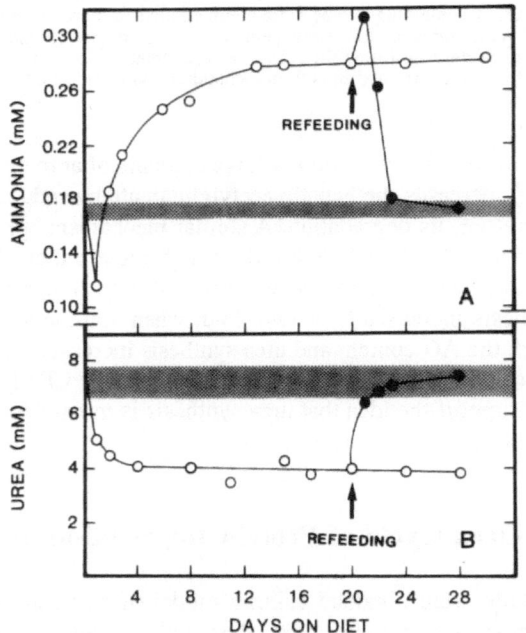

Figure 4. Effects of protein–free diet and of refeeding the standard diet on blood ammonia and serum urea levels. Rats were fed standard or protein–free diet. After 20 days on protein–free diet (indicated by arrows) a group of these rats was transferred to standard diet. Blood was taken from the tail vein and blood ammonia and serum urea levels were determined. Values are the means of duplicate samples from 6–9 rats. From (25).

Urea excretion decreased continuously in rats fed the protein–free diet until it was only 2–3% of controls. By the 4th day after refeeding the standard diet, urea excretion reached the values of controls (Fig. 5). In contrast, the activity of carbamylphosphate synthetase, which decreased to $\approx$ 42% of controls in rats on protein–free diet, was not increased after 4 days of refeeding the standard diet (Fig. 5) and it took about one month to return to control values. This clearly indicates that also under these conditions the amount of CPS is not involved in the control of urea synthesis.

In contrast, there is an excellent correlation between the time–course of the recovery

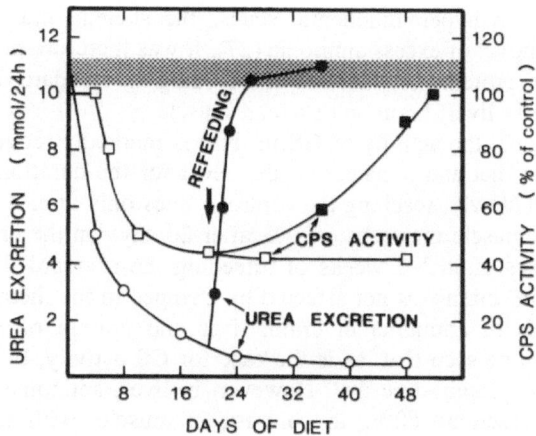

Figure 5. Effects of protein–free diet and of refeeding the standard diet on urea excretion and carbamylphosphate synthetase activity in liver. Rats were fed protein–free diet for up to 50 days. After 20 days on this diet (indicated by arrow) a group of these rats were refed the standard diet. Values are the mean of duplicate samples from five rats. Values ± SD of urea excretion in controls are represented by the stipped area. Values for protein–free diet are given by the open symbols and those for refeeding by the closed symbols. From (25).

of urea synthesis and excretion (Fig. 5) and that of the recovery of hepatic acetylglutamate (Fig. 6).

These results, together with those obtained with the ammonia diet (shown in the previous section), clearly indicate that the synthesis of urea is mainly controlled by the hepatic content of acetylglutamate and that carbamylphosphate synthetase as well as the other enzymes of the urea cycle are in a large excess usually, and also under the different hyperammonemic situations shown here. This is in agreement with the data shown for normal conditions in (26).

5. Effects of Protein Deprivation and Refeeding on Glutamine Synthetase in Several Tissues

Tissues such as brain and skeletal muscle, being devoid of an effective urea cycle, rely on glutamine synthesis via glutamine synthetase (GS) for the removal of excess ammonia. It

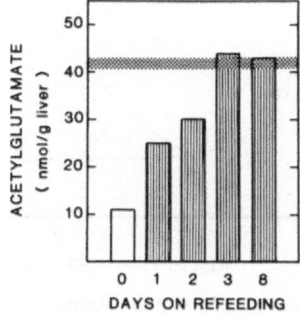

Figure 6. Effect of protein–free diet and of refeeding the standard diet on hepatic acetylglutamate content. Rats were fed protein–free diet for 20 days. Then a group of them was refed the standard diet. Groups of rats were killed at the indicated times and hepatic acetylglutamate was determined. Values are the mean of triplicate samples from four rats. Values ± SD for controls are represented by the dotted area. From (25).

7

has been reported that, in hyperammonemic states, the skeletal muscle could play and important role in the removal of excess ammonia (27). It was therefore of interest to evaluate the effects of feeding the protein–free diet and of refeeding the standard diet on the activities of glutamine synthetase in liver, brain and skeletal muscle.

As shown in Fig. 7, the activity of GS in liver is markedly reduced (80%) after 20 days on the protein–free diet and remained at this level for the duration of the diet; upon refeeding, GS recovered slowly, reaching the control values only after 4 weeks. In contrast, glutamine synthetase in muscle increased ≈ 70% after 20 days on the protein–free diet and returned to control values after 2–3 weeks of refeeding the standard diet. The activity of glutamine synthetase in brain was not affected by changes in the dietary protein (Fig. 7). Fig. 8 shows the levels of ammonia in brain, liver and muscle of rats fed control or protein–free diets. It can be seen that, as is the case for GS activity, brain ammonia levels were not affected by the protein–free diet. However, in liver, ammonia increased by 84% while GS activity decreased by 80%. In contrast, in muscle, with increased glutamine synthetase activity, there was a decrease in ammonia content. It seems therefore, that in these tissues the ammonia concentration is related to the activity of glutamine synthetase. It has recently been shown that portacaval anastomosis in rats also induces a similar pattern of changes in glutamine synthetase activities, with increased activity in muscle, decreased activity in liver and no change in brain (28). It is interesting that two different models of hyperammonemia lead to similar changes in GS activities in different tissues.

These results suggest that in hyperammonemic conditions, glutamine synthesis in muscle could be an important alternative site for ammonia detoxification.

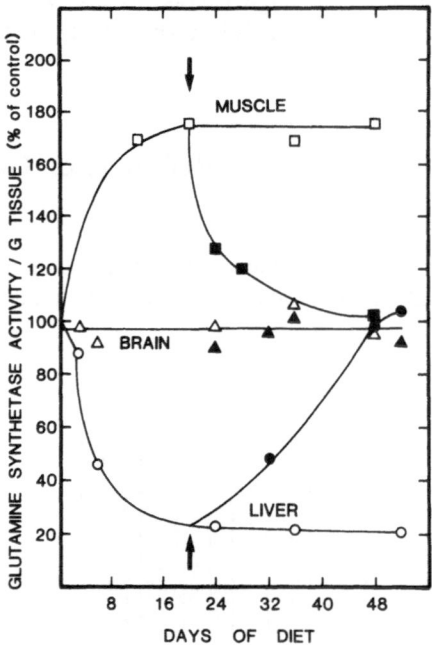

Figure 7. Effect of protein–free diet and of refeeding the standard diet on glutamine synthetase in liver, brain and muscle. Rats were fed as described in the legend to Fig 4. Values (obtained as units/g tissue) are expressed as percentages of control and are the mean of triplicate samples from five rats. Values for the protein–free diet are given by the open symbols and those for refeeding by the closed symbols. From (25).

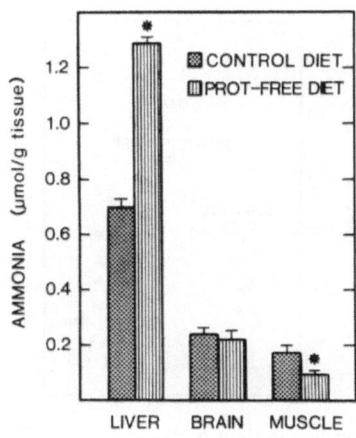

Figure 8. Effect of protein-free diet on ammonia levels in liver, brain and muscle. Rats were fed standard or protein-free diet for 20 days. Values are the means ± SD of duplicate samples from five animals. Data taken from (25).

6. Treatment of Hyperammonemia with Carbamylglutamate in Rats

Clinical treatments of hyperammonemic states are usually directed towards decreasing the blood ammonia levels by reducing its intestinal formation or its transport, by colonic acidification (29). It is also customary to restrict the patients to a low protein diet. As mentioned above, a protein-free diet causes a paradoxical increase in blood ammonia (22, 25), due to the consumption of body proteins through increased proteolysis. Moreover, reduction of the dietary protein results in a decrease in liver of the urea cycle enzymes, including CPS, and of acetylglutamate, its physiological activator, thus accentuating the danger of ammonia intoxication in these patients.

The above results indicate that the capacity for ammonia detoxification and urea synthesis depends mainly on the hepatic content of acetylglutamate. It was therefore of interest to assess whether increasing the activation of CPS could normalize blood ammonia levels in rats fed a protein-free diet. Administration of acetylglutamate is precluded because it is rapidly degraded by a deacylase present in the cytosol (23) but there is another activator of CPS, carbamylglutamate (7, 8), which is not a substrate for this deacylase, is metabolically stable and can reach the mitochondria and activate CPS. Furthermore, carbamylglutamate has been shown to be nontoxic when used even for lengthy periods in human patients (30).

Groups of rats were fed standard or protein-free diets; for some groups, carbamylglutamate was administered (1 mM) in the drinking water (31). To assess if carbamylglutamate is able to reach the mitochondria, we determined the content of activators of CPS in mitochondria isolated from rats of the different groups. Hepatic acetylglutamate contents were 846 ± 265 pmol/mg and 87 ± 32 pmol/mg mitochondrial protein for rats fed standard and protein-free diets, respectively. Livers of rats fed the protein-free diet and ingesting carbamylglutamate contained both acetylglutamate and carbamylglutamate. Their capacity for activating CPS was equivalent to that of 547 ± 231 pmol acetylglutamate/mg mitochondrial protein. This clearly indicates that ingested carbamylglutamate does reach the liver where it activates CPS.

As shown in Fig. 9, ingestion of carbamylglutamate is able to normalize blood ammonia levels in rats fed the protein-free diet. Moreover, the excess ammonia is eliminated as urea (Fig. 10).

These results suggest that oral administration of carbamylglutamate might be useful

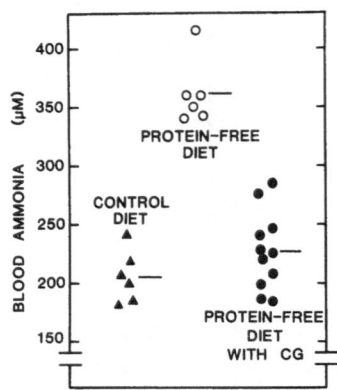

Figure 9. Effect of carbamylglutamate ingestion on blood ammonia levels. Rats were fed standard or protein–free diet for 20 days. Carbamylglutamate was administered in the drinking water (1 mM) to a group of six rats fed the protein–free diet. Samples were then taken from the tail vein once a week. Values are the mean of triplicate samples from at least three rats. From (31).

in the treatment of hyperammonemic states, including CPS deficiencies and in the more common pathologic instances of cirrhosis. It has been reported that CPS activity is reduced in cirrhotic human liver biopsy specimens (15, 16). Moreover, diets with reduced protein content are usually prescribed for patients with cirrhosis, thus further decreasing the hepatic content of CPS and of acetylglutamate. Under these conditions, activation of the remaining CPS by oral administration of carbamylglutamate could be useful in these patients to normalize blood ammonia levels.

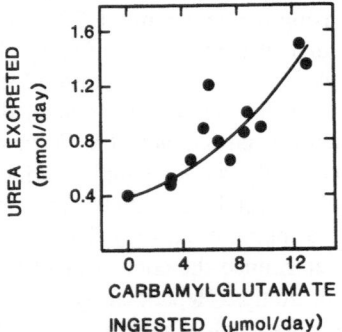

Figure 10. Effect of carbaylglutamate ingestion on urea excretion. Rats were fed the protein–free diet with or without carbamylglutamate in drinking water for 20 days. The rats were then housed in metabolic cages, and urine was collected daily for periods of four consecutive days. Intake of carbamylglutamate solution was also measured for each rat for the same periods of time. Each value is the mean of both the urea excreted and the carbamylglutamate ingested for a given rat over 4 consecutive days. The value for the protein–free group (no carbamylglutamate ingested) is the mean of daily urea xcretion of three different rats during two periods of 4 consecutive days. The group ingesting carbamylglutamate consisted of six rats. From (31).

In conclusion, ammonia detoxication in ureotelic animals is usually controlled mainly by acetylglutamate. Analogs of acetylglutamate such as carbamylglutamate may be useful for therapy in hyperammonemia states, including liver cirrhosis.

REFERENCES

1. Hahn, M., Massen, O., Nenchi, M. and Pavlov, I., 1893, Die Eck'sche fistel zwischen der unteren hohlvene und der pfortader und ihre folgen für den organisms. Arch. Exp. Path. Pharmak. **32**:161–173.

2. Cooper, A. J. L., 1990, Ammonia metabolism in normal and portacaval–shunted rats. Adv. Exp. Med. Biol. 272, 23–46.

3. Cooper, A. J. L. and Plum, F., 1987, Biochemistry and physiology of brain ammonia. Physiol. Rev. 67:440–519.

4. Tokatli, A., Coskun, T., and Özalp, I., 1991, Fifteen years' experience with 212 hyperammonemic cases at a metabolic unit. J. Inher. Metab. Dis. **14**:698–706.

5. Krebs, H. A., and Henseleit, K. J., Physiol. Chem. **210**:33.

6. Krebs, H. A., 1976, The discovery of the ornithine cycle. In: The Urea Cycle Grisolía, S., Báguena, R. and Mayor, F., eds) pp 1–12. John Wiley and Sons., New York.

7. Cohen, P. P., and Grisolía, S., 1948, The intermediate role of carbamyl–L–glutamic acid in citrulline synthesis.J. Biol. Chem. 174:389–39.

8. Grisolía, S., and Cohen, P. P., 1952, The catalytic role of carbamyl–L–glutamic acid in citrulline synthesis. J., Biol. Chem. **198**:561–571.

9. Schimke, R. T., 1962, Adaptive characteristics of urea cycle enzymes in the rat. J. Biol. Chem. **237**:459–468.

10. Schimke, R. T., 1963, Studies on the factors affecting the levels of urea cycle enzymes in the rat. J. Biol. Chem. **238**:1012–1018.

11. Aebi, H., 1976, Coordinated changes in enzymes of the ornithine cycle and response to dietary conditions. In: The Urea Cycle (Grisolía, S., Báguena, R., and Mayor, F., eds) pp 275–299. John Wiley and Sons, New York.

12. Shigesada, K., and Tatibana, M., 1971, N–acetylglutamate synthetase from rat liver mitochondria. J. Biol. Chem. **246**: 5588–5595.

13. Shigesada, K., Aoyagi, K., and Tatibana, M., 1978, Role of acetylglutamate in ureotelism. Variations in acetylglutamate level and its possible significance in control of urea synthesis in mammalian liver. Eur. J. Biochem. **85**:385–391.

14. Stewart, P. M., and Walser, M., 1980, Short term regulation of ureagenesis. J. Biol. Chem. **255**:5270–5280

15. Khatra, B. S., Smith, R. B., Millikan, W. J., Sewell, C. W., Warren, W. D., and Rudman, D., 1974, Activities of Krebs–Henseleit enzymes in normal and cirrhotic human liver. J., Lab. Clin. Med. **84**:708–715.

16. Ugarte, G., Pino, M. E., Valenzuela, J., and Lorca, F., 1963, Urea cycle abnormalities in patients in endogenous hepatic coma. Gastroenterology **45**:182–188.

17. Fujii, T., Kohno, M. and Hirayama, C., 1992, Metabolism of ^{15}N–ammonia in patients with cirrhosis: a three–compartmental analysis. Hepatology **16**:347–352.

18. Müller, W. A., Faloona, G. R., Aguilar–Parada, E., and Unger, R. H., 1970, N., Engl. J. Med. 283 109–115.

19. Snodgrass, P. J., Lin, R. C., Müller, W. A., and Aoki, T. T., 1978, Induction of urea cycle enzymes of rat liver by glucagon. J. Biol. Chem. **253**:2748–2753.

20. Ryall, J. C., Quantz, M. A., and Shore, G. C., (1986, Rat liver and intestinal mucosa differ in the developmental pattern and hormonal regulation of carbamoylphosphate synthetase I and ornithine carbamoyltransferase gene expression. Eur. J. Biochem. **156**:453–458.

21. Azorín, I., Miñana, M. D., Felipo, V. and Grisolía, S., 1989, A simple animal model of hyperammonemia. Hepatology **10**:11–314.

22. Felipo, V., Miñana, M. D., and Grisolía, S., 1988, Long–term ingestion of ammonium increases acetylglutamate and urea levels without affecting the amount of carbamoyl–phosphate synthase. Eur. J., Biochem. **176**: 567–571.

23. Reglero, A., Rivas, J., Mendelson, J., Wallace, R., and Grisolía, S., 1977, Deacylation and transacetylation of acetylglutamate and acetylornithine in rat liver. FEBS Lett. **81**:13–17.

24. Morita, T., Mori, M., and Tatibana, M., 1982, Regulation of N–acetyl–L–glutamate degradation in mammalian liver. J., Biochem. (Tokyo) **91**:563–569.

25. Felipo, V., Miñana, M. D., and Grisolía, S., 1992, Control of urea synthesis and ammonia

utilization in protein deprivation and refeeding. Arch. Biochim. Biophys.285: 351–356.

26. Alonso, E., Girbés, J., García-España, A., and Rubio, V., 1989, Changes in urea cycle–related metabolites in the mouse after combined administration of valproic acid and an amino acid load. Arch. Biochim. Biophys. 272 267–273.

27. Ganda, O. P., and Ruderman, N. B., 1976, Muscle nitrogen metabolism in chronic hepatic insufficiency. Metabolism 25:427–435.

28. Girard, G., and Butterworth, R. R., 1992, Effect of portacaval anastomosis on glutamine synthetase activities in liver, brain and skeletal muscle. Dig. Dis. Sci. 37:1121–1126.

29. Uribe, M., 1990, Treatment of portal systemic encephalopathy: The old and new treatments. In: Cirrhosis, Hepatic Encephalopathy and Ammonium Toxicity (Grisolía, S., Felipo, V., and Miñana, M. D.,), eds Adv. Exp. Med. Biol. 272:235–254.

30. Brown, R., Manning; R., Delp, M., and Grisolía, S., 1958, Treatment of hepatocerebral intoxication. Lancet 1:591–592.

31. Grau, E., Felipo, V., Miñana, M. D., and Grisolía, S., 1992, Treatment of hyperammonemia with carbamylglutamate in rats. Hepatology 15:46–448.

32. Grisolía, S., Felipo, V., Miñana, M. D., Costell, M., and O'Connor, J. E., 1989, Aspects of the participation of the urea cycle in the protection against ammonium toxicity. Bull. Mol. Biol. Med. 14:191–210.

Brain Metabolism in Hepatic Encephalopathy and Hyperammonemia

Richard A. Hawkins and Anke M. Mans

1. Introduction

In the United States of America liver failure is the sixth leading cause of death in persons aged 25–65 years. Liver failure may occur within days as a result of fulminant hepatic necrosis or over many years in chronic conditions such as alcoholic fatty liver or cirrhosis. When the liver fails, or when blood is shunted around a cirrhotic liver directly from the intestines into the systemic circulation, brain function deteriorates: a disorder known as hepatic encephalopathy [1, 18, 36,47, 53]. This syndrome is manifest by signs that range from a rapidly developing sequence of delirium, convulsions and coma in fulminant hepatic necrosis to a more gradually developing intellectual impairment that may lead to stupor and coma in patients with chronic liver disease. The latter form is more prevalent and may affect several millions of people to some degree [34, 44].

The reversibility of the signs in some patients and the absence of damage to neurons suggest that the encephalopathy has a metabolic cause. Brain function is compromised even in the milder stages of portalsystemic encephalopathy [11, 12, 42], and becomes more sensitive to drugs and metabolic disturbances that would have no adverse consequences in normal individuals [18, 36, 53].

The mechanism by which liver failure leads to a disturbance in brain function remains obscure, in spite of much information that is now available about the associated metabolic changes. The degree of encephalopathy in humans and rats is reflected by a measurable reduction in the rate of the use of glucose, the principal source of energy, throughout the brain (reviewed in [16]). This diminished rate of energy consumption is most likely explained by a decrement in the activity of brain cells. Because of this relationship a reduction in the rate of brain energy consumption is a convenient measure of diminished cerebral function in experimental animals. There are also various other metabolic changes that may be of etiologic importance. In rats with a portacaval shunt, a model of chronic liver disease which causes liver atrophy, the following changes are found: an increase in plasma and brain ammonia [27], a pronounced change in the plasma amino acid spectrum (decreases in the branched–chain amino acids and threonine and increases in the aromatic amino acids) [19, 30], a large rise in the brain content of aromatic amino acids and glutamine [19, 27], increased permeability of the blood–brain barrier to neutral amino acids [19, 29, 30, 43], and increased brain content of serotonin (and its metabolite, 5–HIAA) and norepinephrine [32]. The relationship between these diverse metabolic alterations and encephalopathy remains to be elucidated.

The onset of hepatic encephalopathy occurs soon after portacaval shunting [31]. Brain

Department of Physiology and Biophysics, University of Health Sciences/The Chicago Medical School, 3333 Green Bay Road, North Chicago, IL 60064, USA

Cirrhosis, Hyperammonemia, and Hepatic Encephalopathy,
Edited by S. Grisolia and V. Felipo, Plenum Press, New York, 1994

energy metabolism starts its downward course within six hours and is maximally depressed within one or two days [10, 31]. This reduced rate of energy consumption is maintained for at least several months thereafter [31]. Most of the other abnormalities that are characteristic of the condition are also established by one day, with the exception of the reduction in plasma branched–chain amino acids and threonine, and the increase in brain norepinephrine, which take about two days. The latter changes are, therefore, less likely to be etiologically important.

Ammonia has long been suspected to be an important factor in the cerebral dysfunction of hepatic encephalopathy and other diseases in which pronounced hyperammonemia occurs [6, 9, 36]. Portacaval shunting raises the plasma ammonia levels from a normal value of about 50 nmol/ml to 200–800 nmol/ml. These high concentrations of arterial ammonia seem to originate from intestinal metabolism of glutamine as well as through the production of ammonia by intestinal bacteria [48, 51]. Regardless of the origin of the ammonia, reducing the degree of hyperammonemia is one of the primary objectives of therapy. The relationship between hyperammonemia and encephalopathy, however, is puzzling because the blood ammonia levels do not always correlate with the severity of the neurologic signs [6, 9, 26, 46, 53].

2. Hyperammonia Alone Causes Similar Metabolic Abnormalities

Recent experiments showed thtat hyperammonemia alone could cause several of the metabolic abnormalities associated with portacaval shunting [22, 23]. In these experiments hyperammonemia was induced by intraperitoneal injections of urease rather than through shunting of hepatic portal blood. After two days there was decreased brain glucose consumption, increased transport of tryptophan across the blood–brain barrier, and increased glutamine and aromatic amino acids in brain: the same pattern of metabolic changes as that observed in portacaval shunted rats. These experiments showed that ammonia was indeed a major factor in producing these changes.

3. Central Position of Glutamine Synthetase

In the experiments described above on artificially induced hyperammonemia, the decrease in brain energy consumption, and presumably brain function, correlated more closely with increased brain glutamine than with increased plasma ammonia. This was similar to the situation previously described in humans with portalsystemic encephalopathy in which encephalopathy was most closely related with glutamine or one of its metabolites, alpha–ketoglutaramate [17, 49]. The close relationship indicated that glutamine synthesis might be an important connection in the sequence between hyperammonemia and brain dysfunction.

Further experiments showed that glutamine synthesis is indeed an essential step in the adverse response to ammonia. In the absence of net glutamine synthesis hyperammonemia caused no detectable abnormalities [14]. In these experiments glutamine accumulation and hyperammonemia were separated as causes of encephalopathy by raising plasma ammonia levels with a small dose of methionine sulfoximine, an inhibitor of glutamine synthetase. This treatment caused hyperammonemia without an increase in brain glutamine content. These hyperammonemic rats, with plasma and brain ammonia levels equivalent to those associated with decreased brain consumption in portacaval shunted rats, behaved normally during the two days of study. There was no depression of cerebral energy consumption, and the concentrations of key intermediary metabolites and high energy phosphates in the brain were normal. Neutral amino acid transport (tryptophan and leucine) and the brain content of aromatic amino acids were unchanged. The data suggested that ammonia is benign at concentrations of 1mM or less if it is not converted to glutamine. Thus, the deleterious effects of chronic hyperammonemia seem to begin with the synthesis of glutamine. The glutamine synthetase reaction, long considered to be a mechanism of ammonia detoxification [6, 9, 18, 25, 36, 50, 53], may in fact be the first step causing the metabolic abnormalities.

4. Amelioration of Metabolic Abnormalities

We tested whether the metabolic signs of portalsystemic encephalopathy in rats with portacaval shunts could be prevented or diminished by inhibiting the metabolism of ammonia by glutamine synthetase in the brain, using methionine sulphoximine. We showed that the onset of metabolic signs of encephalopathy caused by portacaval shunting could be prevented to some degree, and that established metabolic abnormalities could be partly reversed, by reducing glutamine synthetase activity in the brain [15]. To do this, small dose of methionine sulphoximine, sufficient to partially inhibit brain glutamine synthetase, was given to rats either at the time of portacaval shunting or three to four weeks later. The effects on several characteristic cerebral metabolic abnormalities produced by portacaval shunting were measured one to three days after injection of the inhibitor. All untreated portacaval shunted rats had elevated plasma and brain ammonia concentrations, increased brain glutamine and tryptophan content, decreased brain glucose consumption and increased permeability of the blood–brain barrier to tryptophan. In all treated rats brain glutamine content was normalized, indicating inhibition of glutamine synthesis. One day after shunting and methionine sulphoximine administration, glucose consumption, tryptophan transport and brain tryptophan content remained near normal. In the three–to four–week shunted rats, studied one to three days after methionine sulphoximine administration, the effect was less pronounced; brain glucose consumption and tryptophan content were partially normalized, but tryptophan transport was unaffected. The results agree with the earlier conclusion that glutamine synthesis is an essential step in the development of cerebral metabolic abnormalities in hyperammonemic states, and suggest that this step deserves further attention.

5. Possible Mechanisms

5.1. Energy Reduction

It has been proposed that there may be an energy and metabolite drain, caused by abnormally high rates of glutamine synthesis. It is also conceivable that ammonia interferes with brain energy production by inhibition of the alpha–oxoglutarate dehydrogenase complex or the mitochondrial malate–aspartate shuttle (see [9] for a summary). These hypotheses do not seem to offer an adequate explanation for cerebral dysfunction in chronic hyperammonemia. Such interference with energy production would cause predictable and substantial alterations in the concentrations of several intermediary metabolites, but these changes have not been observed in vivo (reviewed in [16]). Furthermore, while a temporary deficiency in energy production could interfere with cell function, a sustained deficit would result in cell death. The neuropathology of hepatic encephalopathy shows no such evidence. Thus, the finding of reduced cerebral metabolic rates of glucose and oxygen in hepatic encephalopathy must be explained by a reduction in the demand, i.e., the level of work.

The hyperammonemia produced by portacaval shunting does not impose much of a metabolic burden on brain. Even when lethal doses of exogenous ammonia were given, a reduction in phosphocreatine and ATP levels seemed to be a secondary rather than a causal phenomenon [45]. Although the metabolic demand made by increased glutamine synthesis in chronically hyperammonemic rats forms a relatively small percentage of total brain metabolism, it falls entirely on astrocytes which comprise perhaps one third of the total cell volume in brain. While astrocytes do show morphological changes, which may be compensatory to this process, there are no signs of astrocyte cell death, as might be expected in a situation of long–lasting energy failure [8, 35, 52]. It is possible that the sustained hyperactivity may disrupt other activities of the astrocytes, such as regulating neurotransmitter uptake and ion homeostasis, and influencing blood–brain barrier transport of essential nutrients. These possibilities remain to be tested.

5.2. Inhibition of Glutaminase

It has been suggested that ammonia may disturb cerebral function by inhibiting glutaminase [3, 4, 33] thereby disrupting the supply of glutamate for glutamatergic neurons [5, 7, 13]. Our experiments do not support this hypothesis. Methionine sufoximine caused hyperammonemia, but the brain content of glutamine and glutamate remained near normal. If there were inhibition of glutaminase by ammonia the glutamate concentration would have been expected to decrease. It is conceivable that the lack of a decrease could be explained if glutaminase was inhibited by ammonia to the same degree that glutamine synthetase was inhibited by methionine sulfoximine. However, in that circumstance production of glutamate from glutamine would have been reduced even more.

5.3. Ammonium Ion

There are at least two known effects of NH_4^+ on neuronal membranes that could disturb brain function. NH_4^+ can interfere with the generation of inhibitory postsynaptic potentials by inhibiting the Cl^- pump. This occurs at ammonia concentrations similar to those reported here and is maximal at a concentration of 1 mM [38, 40]. At higher concentrations (greater than 2 mM) NH_4^+ can depolarize neurons and interfere with synaptic transmission [38, 40].

In our experiments using methionine sulphoximine the brain concentrations of ammonia were high enough to inhibit the Cl^- pump and, at least in principle, to cause neurons to become more easily excited [37]. However there were no detectable changes in brain glucose consumption or in the concentrations of high–energy intermediates [14]. It is known that neurons adapt to the chronic presence of NH_4^+ and it is possible that the Cl^- pump had recovered by twenty four hours [39]. Therefore, while it remains conceivable that NH_4^+ directly interferes with nerve cell function in hyperammonemic diseases, the effects may be too subtle to be detected by the techniques used.

5.4. Stimulation of Blood–brain Barrier Transport

It is well established that hyperammonemia, whether caused by portacaval shunting or by artificial means, leads to a greater activity of the carrier system that transports neutral amino acids across the blood–brain barrier, as well as to a rise in the brain content of aromatic amino acids [2, 19, 20, 23, 24, 28, 29, 30]. The permeability of the blood– brain barrier to neutral amino acids and the accumulation of aromatic amino acids are both closely correlated with brain glutamine content [21, 22, 23]. Moreover the elevated rate of transport and the increased accumulation of neutral amino acids caused by hyperammonemia or by portacaval shunting can be reduced by treatment with methionine sulfoximine [24, 41]. Our experiments, in agreement with previous observations, show conclusively that sustained hyperammonemia in the absence of net glutamine synthesis has no effect on either neutral amino acid transport or the accumulation of aromatic amino acids in brain. It may, therefore, be concluded that glutamine synthesis is linked to the stimulation of neutral amino acid transport caused by hyperammonemia as well as to the decrease in cerebral energy metabolism [22, 23]. After portacaval shunting the change in neutral amino acid transport occurs very early, beginning within six hours, and is paralleled by a substantial decrease in cerebral glucose consumption [31]. The relationship between the changes in the permeability of the blood–brain barrier and cerebral dysfunction needs to be clarified.

6. Concluding Comments

It is now evident that hyperammonemia disturbs cerebral function and is an important trigger in the sequence of events leading to hepatic encephalopathy. The absence of a toxic response to hyperammonemia when it is created by inhibiting glutamine synthetase, indicates that ammonia itself is relatively innocuous (at concentrations below 1mM). It is when ammonia is metabolized that an adverse response is initiated. This does not mean that glutamine itself is toxic, there are many other metabolites and cellular precesses to be considered. Nevertheless, the observation that the metabolism of ammonia plays an essential role in causing cerebral dysfunction has important implications for diseases in which hyperammonemia is a characteristic feature like hepatic encephalopathy and some inborn errors of metabolism. Certainly, attention should be focused on the glutamine synthetase reaction and its products in the search for the mechanisim of ammonia toxicity.

ACKNOWLEDGMENT: The authors' work was supported by NIH grant NS–16389.

References

1. Adams, R. D., and Foley, J. M., 1953, The neurological disorder associated with liver disease, Res. Publ. Assoc. Nerve. Ment. Dis. 32:198–237.
2. Bachmann, C., and Colombo, J. P., 1984, Increase of tryptophan and 5–hydroxyindole acetic acid in the brain of ornithine carbamoyltransferase deficient sparse–fur mice, Pediatr. Res. 18:372–375.
3. Benjamin, A. M., 1881, Control of glutaminase activity in rat brain cortex in vitro, influence of glutamate, phosphate, ammonium, calcium and hydrogen ions, Brain Res. 208:363–377.
4. Bradford, H. F., and Ward, H. K., 1975, Glutamine as a metabolic substrate for isolated nerve-endings: inhibition by ammonium ions, Biochem. Soc. Trans. 3:1223–1226.
5. Bradford, H. F., Ward, H. K., and Thomas, A. J., 1978, Glutamine –A major substrate for nerve endings, J. Neurochem. 30:1453–1459.
6. Butterwork, R. F., Giguere, J. F., Michaud, J., Lavoie, J., and Pomier–Layrargues, G., 1987, Ammonia: Key factor in the pathogenesis of hepatic encephalopathy, Neurochem. Pathol. 6:1–12.
7. Butterworth, R. F., Lavoie, J., Giguere, J. F., Layrargues, G. P., and Bergeron, M., 1987, Cerebral GABA–ergic and glutamatergic function in hepatic encephalopathy, Neurochem. Pathol. 6:131–144.
8. Cavanagh, J. B., and Kyu, M. H., 1971, Type II Alzheimer change experimentally produced in astrocytes in the rat, J.Neurol. Sci. 12:63–75.
9. Cooper, A. J., and Plum, F., 1987, Biochemistry and physiology of brain ammonia, Physiol. Rev. 67:440–519.
10. DeJoseph, M. R., and Hawkins R. A., 1991, Glucose consumption decreases throughout the brain only hours after portacaval shunting, Am. J. Physiol. 260:E613–E619.
11. Elsass, P., Lund, Y., and Ranek, L., 1978, Encephalopathy in patients with cirrhosis of the liver. A neuro–psychological study, Scand. J. Gastroenterol. 13:241–247.
12. Gilbestadt, S. J., Gilberstadt, H., Zieve, L., Buegel, B., Collier Jr, R. O., and McClain, C. J., 1980,

Psychomotor performance defects in cirrhotic patients without overt encephalopathy, Arch. Intern. Med. **140**:519–521.

13. Hamberger, A., Hedquist, B., and Nystrom, B., 1979, Ammonium ion inhibition of evoked release of endogenous glutamate from hippocampal slices, J. Neurochem. **33**:1295–1302.

14. Hawkins, R. A., and Jessy, J., 1991, Hyperammonemia does not impair brain function in the absence of glutamine hesis, Biochem. J. **277**:697–703.

15. Hawkins, R. A., Jessy, J., Mans A. M., and De Joseph, M. R., 1992, Effect of reducing brain glutamine synthesis on metabolic signs of hepatic encephalopathy, J Neurochem. In press.

16. Hawkins, R. A., and Mans, A. M., 1989, Brain energy metabolism in hepatic encephalopathy. In: Hepatic Encephalopathy, edited by Butterworth, R. F., and Pomier–Layrargues, G., 1989, Clifton, NJ: Humana Press Inc., p. 159–176.

17. Hourani, B. T., Hamlin, E. M., and Reynolds, T. B., 1971, Cerebrospinal fluid glutamine as a measure of hepatic encephalopathy, Arch. Intern. Med. **127**:1033–1036.

18. Hoyumpa Jr., A. M., Desmond, P. V., Avant, G. R., Roberts, R. K.,and Schenker, S., 1979, Hepatic encephalopathy, Gastroenterology, **76**:184–195.

19. James, J. H., Escourrou, J., and Fischer, J. E., 1978, Blood–brain neutral amino acid transport activity is increased after portacaval anastomosis, Science **200**:1395–1397.

20. James, J. H., Hodgman, J. M., Funovics, J. M., and Fischer, J. E., 1976, Alterations in brain octopamine and brain tyrosine following portacaval anastomosis in rats. J. Neurochem. **27**: 223–227.

21. Jeppsson, B., James, J. H., Edwards, L. L., and Fischer, J. E., 1985, Relationship of brain glutamine and brain neutral amino acid concentrations after portacaval anastomosis in rats, Eur. J. Clin. Invest. **15**:179–187.

22. Jessy, J., DeJoseph, M. R., and Hawkins, R. A., 1991, Hyperammonemia depresses glucose consumption throughout brain. Biochem. J. **277**:693–696.

23. Jessy, J., Mans, A. M., DeJoseph, M. R., and Hawkins, R. A., 1990, Hyperammonemia causes man of the changes found after portacaval shunting, Biochem. J. **272**:311–317.

24. Jonung, T., Rigotti, P., Jeppsson, B., James, J. H., Peters J. C., and Fischer, J. E., 1984, Methionine sulfoximine prevents the accumulation of large neutral amino acids in brain of hyperammonemic rats, J. Surg. Res. **36**:349–353.

25. Krebs, H. A., 1936, Metabolism of amino acids, IV, The synthesis of glutamine animal tissue, Biochem. J. **29**:1951–1969.

26. Lockwood, A. H., 1987, Metabolic encephalopathies: opportunities and challenges. J. Cereb. Blood Flow Metab. **7**:523–526.

27. Mans, A. M., Biebuyck, J. F., Davis, D.W., and Hawkins, R. A., 1984, Portacaval anastomosis: brain and plasma metabolite abormalities and the effect of nutritional therapy, J. Neurochem. **43**:697–705.

28. Mans, A. M., Biebuyck, J. F., and Hawkins, R. A., 1983, Ammonia selectively stimulates neutral amino acid transport across blood–brain barrier, Am. J. Physiol. **245**:C74–C77.

29. Mans, A. M., Biebuyck, J. F., Saunders, S. J., Kirsch, R. E., and Hawkins, R. A., 1979, Tryp tophan transport across the blood–brain barrier during acute hepatic failure, J. Neurochem. **33**:409–418.

30. Mans, A. M., Biebuyck, J. F., Shelly, K., and Hawkins, R. A., 1982, Regional blood–brain barrier permeability to amino acids after portacaval anastomosis, J. Neurochem. **38**:705–717.

31. Mans, A. M., DeJoseph, M. R., Davis, D. W., Viña, J. R., and Hawkins, R. A., 1990, Early establishment of cerebral dysfunction after portacaval shunting, Am. J. Physiol. **258**:E104–E110.

32. Mans, A. M., and Hawkins, R. A., 1986, Brain monoamines after portacaval anastomosis, Metab. Brain Dis. **1**:45–52.

33. Matheson, D. F., and Van den Berg, C. J., 1975, Ammonia and brain glutamine: inhibition of glutamine degradation by ammonia, Biochem. Soc. Trans. **3**:525–528.

34. Mendenhall, D. L., 1981, Alcoholic Hepatitis. Clin. Gastroenterol. **10**:417–441.

35. Norenberg, M. D., 1979, The distribution of glutamine synthetase in the rat central nervous system, J. Histochem. Cytochem. **27**:756–762.

36. Plum, F., and Hindfelt, B., 1976, The neurological complications of liver disease. In: Metabolic and Deficiency Diseases of the Nervous System. Part I, edited by Vinken, P. J., Bruyn G. W., and Klawans, H. L., 1976, New York: American Elsevier Publishing Co. Inc. p. 349–377.

37. Raabe, W., 1982, Ammonia and postsynaptic inhibition in cat motor cortex. In: Physiology and Pharmacology of Epileptogenic Phenomena, edited by Klee, M. R., Lux, H. D., and Speckmann, E. J., New York: Raven Press, p. 73–80.

38. Raabe, W., 1989, Ammonium decreases excitatory synaptic transmission in cat spinal cord in vivo, J. Neurophysiol. **62**:1461–1473.

39. Raabe, W., 1989, Neurophysiology of ammonia intoxication. In: Hepatic Encephalopathy: Pathophysiology and Treatment, edited by Butterworth, R., and Pomier–Layrargues, G., Clifton, NJ: Humana Press, Inc., p.49–77.

40. Raabe, W., 1991, Effects of NH_4^+ on the function of the CNS, Adv. Exp. Med. Biol. **272**:89–98.

41. Rigotti, P., Jonung, T., Peters, J. C., James, J. H., and Fischer, J. E., 1985, Methionine sulfoximine prevents the accumulation of large neutral amino acids in brain of portacaval-shunted rats, J. Neurochem. **44**:929–933.

42. Rikkers, L., Jenko, P., Rudman, D., and Freides, D., 1978, Subclinical hepatic encephalopathy: detetion, prevalence, and relationship to nitrogen metabolism. Gastroenterology. **75**:462–469.

43. Sarna, G. S., Bradbury, M. W., Cremer, J. E., Lai, J. C., and Teal, H. M., 1979, Brain metabolism and specific transport at the blood–brain barrier after portacaval anastomosis in the rat, Brain Res. **160**:69–83.

44. Scheig, R., 1991, That demon rum. Am. J. Gastroenerol. **86**:150–152.

45. Schenker, S., and Brady, C. E., 1990, Pathogenesis of hepatic encephalopathy. In: Hepatic Encephalopathy: Management with Lactulose and Related Carbohydrates, edited by Conn, H. O., and Bircher, J., East Lansing, MI: Medi–Ed Press, p. 15–30.

46. erlock, S., 1958, Pathogenesis and management of hepatic coma, Am. J. Med. **24**:805–813.

47. Sherlock, S., Summerskill, W. H. J., White, L. P., and Phear, E. A., 1954, Portalsystemic encephalopathy. Neurological complications of liver disease, Lancet **2**:453–457.

48. Soeters, P. B., van Leuwen, P. A. M., and van Berlo, C. L. H., 1989, Nitrogen metabolism in the gut. In: Hepatic Encephalopathy: Management with lactulose and related carbohydrates, edited by Conn, H. O., and Bircher, J., East Lansing, MI: Medi–Ed Press, p. 31–40.

49. Vergara, F., Plum, F., and Duffly, T. E., 1974, ₂-Ketoglutaramate: Increased concentrations in the cerebrospinal fluid of patients in hepatic coma. Science **183**:81–83.

50. Weil–Malherbe, H., 1950, Significance of glutamic acid for the metabolism of nervous tissue, Physiol. Rev. **30**:549–568.

51. Windmueller, H. G., 1984, Metabolism of vascular and luminal glutamine by intestinal mucosa in vivo. In: Glutamine Metabolism in Mammalian tissues, edited by Gayssubgerm, D., and Sies, H., Berlin: Springer–Verlag, p. 61–77.

52. Zamora, A. J., Cavanagh, J. B., and Kyu, M. H., 1973, Ultrastructural responses of the astrocytes to portacaval anastomosis in the rat, J. Neurol. Sci. **18**:25–45.

53. Zieve, L., 1979, Hepatic encephalopathy: summary of present knowledge with an elaboration on recent developments. In: Progress in Liver Diseases, edited by Popper, H., and Schaffner, F., New York: Grune and Stratton, p. 327–341.

Ammonia Metabolism in Mammals: Interorgan Relationships

Arthur J. L. Cooper

1. Introduction

In nature nitrogen can exist in many oxidation states ranging from +5 (nitrate), +3 (nitrite), 0 (N_2) to −3 (NH_3). Many organisms are able to assimilate nitrogen by reduction of compounds containing nitrogen at its higher oxidation states to the −3 oxidation state (ammonia) which in turn is incorporated into amino acids in a reaction requiring removal of two electrons (1). Until relatively recently it was thought that in mammals all enzymatic reactions of nitrogen occur at oxidation states involving =NH, $-NH_2$, or NH_3. In 1978, however, Tannenbaum et al. (2) showed that nitrite and nitrate are generated within the human intestine possibly by bacterial heterotrophic nitrification in the upper part of the intestine. The nitrite/nitrate may be used by other bacteria in the gut, may be involved in formation of N−nitroso compounds or may be reduced (2−4). More recently it has been found that mammals can catalyze formation of NO (nitric oxide, +2 oxidation state) from the ω−nitrogen (i.e. $=NH_2$) of L−arginine in a fascinating reaction that requires a five electron transfer. The free radical NO is involved in the physiological functioning of the mammalian cardiovascular system, immune system and central nervous system (5 and references quoted therein). Nevertheless, it is still apparent that in mammals most nitrogen is ingested, assimilated and excreted in negative oxidation states. Ammonia plays a pivotal role in these processes and is thus a key component of mechanisms used to maintain nitrogen homeostasis. Ammonia is generated and consumed in a large number of reactions within the body; yet despite its central importance ammonia at elevated concentrations is toxic, especially to the central nervous system, and its concentration must be kept low. The physical properties of ammonia, enzymatic reactions in which ammonia is generated or consumed, and the interorgan flux of ammonia and related compounds are reviewed in the chapter.

2. Physical Properties of Ammonia

In aqueous solution ammonia free base (NH_3) is in equilibrium with ammonium ion (NH_4^+). (For convenience except where noted, throughout the chapter ammonia refers to the

Departments of Biochemistry and Neurology, Cornell University Medical College, 1300 York Avenue New York, NY 10021 USA

Cirrhosis, Hyperammonemia, and Hepatic Encephalopathy,
Edited by S. Grisolia and V. Felipo, Plenum Press, New York, 1994

sum of ammonia free base plus ammonium.)   According to the Henderson–Hasselbalch equation:

$$pH = pKa + \log ([NH_3]/[NH_4^+])$$

The pKa for ammonia strongly depends on temperature and ionic strength. For example, at infinite dilution the pKa is 9.24 at 25°C and 8.95 at 35°C; in 2M NH_4NO_3 the pKa is 9.50 at 25°C and 9.05 at 39.9°C (6). It is generally assumed that the pKa of ammonia in blood at 37°C is 9.1 – 9.2 (e.g.7) in agreement with the thermodynamic parameters quoted above. Thus, under normal physiological conditions only about 1% of ammonia is in the form of the free base (NH_3). Ammonia free base is much more soluble than CO_2 in water and lipid (8) and considerably more soluble in plasma than in water, presumably because of its lipophilicity (9). It has been estimated from rapid mixing studies that ammonia concentrations reach equilibrium across the red blood cell membrane in about 100 msec (10).   The kinetics of ammonia uptake into brain have been extensively studied (11). Uptake across the blood–brain barrier (BBB) is largely by diffusion of the free base (11) although a small component may enter by diffusion as ammonium ion (12). Assuming that the intracellular pH of the brain is 7.4 and that of the plasma is 7.1 then one can calculate that the relative concentration of ammonia in brain to that in plasma at equilibrium should be about two. In fact, most, but not all, reports suggest that this ratio is indeed about two (13). However, it is apparent that there is some limitation to diffusion of ammonia across the BBB and under pathophysiological conditions this ratio can change abruptly, suggesting that mixing between blood and brain ammonia pools is not a rapid equilibrium process. For further discussion see reference 13.

3. Enzymes Involved in Ammonia Removal

The most important enzymes involved in ammonia removal in mammals are 1) glutamate dehydrogenase, 2) glutamine synthetase and 3) carbamyl phosphate synthetase I. The glutamate dehydrogenase reaction is freely reversible, but thermodynamically the reaction favors glutamate formation. However, at least in the liver where ammonia is continuously removed it is likely that glutamate dehydrogenase is responsible for the net removal of glutamate.   Braunstein (14) has pointed out that coupling of α–ketoglutarate–utilizing aminotransferases (equ.1) to glutamate dehydrogenase (equ.2) allows ammonia nitrogen to be incorporated or removed from an amino acid:

	L–Amino acid +α–ketoglutarate	⇌	α–keto acid + L–glutamate	(1)
	L–Glutamate + NAD$^+$	⇌	α–ketoglutarate + NADH + NH_4^+	(2)
Net:	L–Amino acid + NAD$^+$	⇌	α–keto acid + NADH + NH_4^+	(3)

Braunstein coined the words "transdeamination" and "transreamination" to describe the forward and reverse directions of reaction 3. For a discussion of the metabolic interactions of glutamate dehydrogenase with aminotransferases see ref. 15.

Glutamine synthetase is present in relatively high concentrations in liver and brain. The enzyme is also present in lower specific activity in skeletal muscle, heart, lung, kidney and other organs. The enzyme exhibits a relatively low Km for ammonia (180 μM (16)) (the concentration of ammonia in physiological fluids is about 100 – 300 μM). Thus, it is probable that glutamine synthetase is the major enzyme involved in removing excess ammonia in extrahepatic organs. Because of its large bulk the skeletal musculature is particularly important in this regard (17). Muscle glutamine synthetase appears to have different properties from that of liver and brain (18). Nevertheless, in both murine and avian species a single gene gives rise to a tissue-specific pattern of expression (19). Glutamine is a particularly abundant amino acid, occuring in millimolar concentrations in rat brain, heart, liver, lung, stomach, small intestine and skeletal muscle (20); like ammonia, glutamine is of pivotal importance in nitrogen homeostasis. Glutamine accounts for about 20% and 67%, respectively, of the amino acid content of human plasma and CSF (21). The widespread occurence of glutamine synthetase probably accounts for the high concentration of glutamine in most body tissues. Glutamine acts as a non-toxic carrier of ammonia, and the amide nitrogen (and to a lesser extent the amine nitrogen) is used in a large number of biosynthetic reactions. These reactions are summarized in Figure 1. Additionally, glutamine is a major energy source of small intestine (22), bone (23), and human diploid fibroblasts (24), and may contribute to energy metabolism in the kidney (25). Finally, glutamine synthetase plays an especially important role in the CNS. In the CNS, glutamine synthetase is primarily located within the astrocytes (26). It is thought that glutamine generated in the astrocytes from the detoxification of ammonia and from the uptake of released neurotransmitter glutamate acts as a source of carbon and nitrogen in the nerve endings to replenish glutamate (27) and possibly GABA (28) released from the nerve terminals. For a review of the occurrence and metabolic importance of glutamine synthetase see ref. 29.

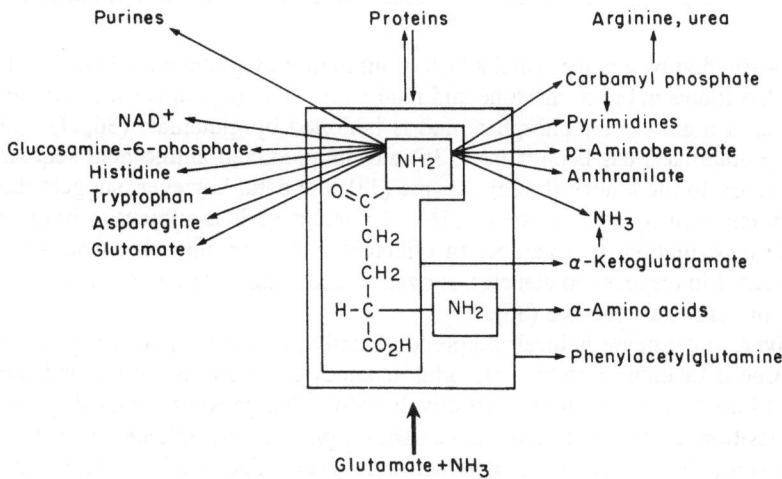

Figure 1. Central role of glutamine in nitrogen homeostasis. The bold arrow emphasizes the fact that glutamine formation is catalyzed by only one known enzyme (glutamine synthetase), whereas glutamine breakdown is catalyzed by numerous enzymes. [Not all reactions shown occur in mammals.] From 30 as adapted from 31. Reproduced with permission.

Carbamyl phosphate synthetase I, the enzyme responsible for introducing ammonia nitrogen into the urea cycle, is found predominantly in the liver wherein it is localized mostly in the periportal hepatocytes. In contrast, glutamine synthetase in liver is localized to a small ring of perivenous hepatocytes (e.g., 32-34). In the fed rat, most of the ammonia nitrogen in the portal vein is incorporated into urea in the liver with only a small percentage incorporated into glutamine (35). The low incorporation into glutamine, despite high activity of glutamine synthetase in the liver, is presumably due to the downstream localization of the enzyme within the sinusoid. Evidently, glutamine synthetase acts as a scavenger for portal vein ammonia and this may have important physiological consequences (see below).

4. Enzymes Involved in Ammonia Formation

Table 1 lists a number of enzymes (or enzyme systems) that are capable of generating ammonia in mammals. The list is not complete but is meant to show the diversity of reactions generating ammonia in the breakdown of nitrogenous compounds.

Quantitatively, the most important ammonia–generating enzymes are the first four on the list, and these will be discussed in the most detail. The major enzyme catalyzing breakdown of glutamine in most tissues is phosphate–activated glutaminase (PAG) of which there are two isozymes. The liver (or L) type is found only in liver and is characterized by a relatively high Km for glutamine, a low Ka for phosphate, a remarkable feed–forward

Table 1. Some enzymes involved in generating ammonia in mammals

Glutaminase L	β–Ureidopropionate(–isobutyrate) deaminase
Glutaminase K	Formiminoglutamate hydrolase
Glutamate dehydrogenase	Transglutaminase
AMP deaminase	Monoamine oxidases
Serine/threonine deaminase	Thyrotropin releasing hormone deamidase
Asparaginase	Guanosine/guanine/adenosine deaminases
Glutamine transaminase L plus ω–amidase	Deoxycytidylate/cytidylate deaminase
Glutamine transaminase K plus ω–amidase	Nucleoside (= (deoxy)cytosine) deaminase
Asparagine transaminase plus ω–amidase	Histidase
Glycine cleavage system	γ–Cystathionase
L– and D–Amino acid oxidases	Porphobilinogen deaminase

(substrate) activation by ammonia and a lack of inhibition by glutamate (36). The kidney (or K) type is also found in brain, intestine and fetal liver; this enzyme has a relatively low Km for glutamine, a high Ka for phosphate and is inhibited by glutamate (36,37). The cDNA of rat liver glutaminase has been cloned (36) and its predicted amino acid sequence shows >80% homology to the kidney (brain) enzyme (38); the data, however, suggest that the two enzymes are encoded by separate genes (36). Glutaminase L is subject to both short–term and long–term regulation by a number of effectors including hormones and ammonia: the activity increases in response to diabetes, starvation and a high protein diet but is not affected by changes in acid–base balance (36).

In liver, glutaminase is localized predominantly in periportal hepatocytes (39). Portal vein glutamine is taken up into the periportal hepatocytes where its carbon and nitrogen are incorporated into glucose and urea, respectively (40). This process is possibly aided by the close juxtapositioning of glutaminase and carbamyl phosphate synthetase I in liver (41).

Glutamate dehydrogenase is most active in liver. The specific activity of the brain enzyme is about 13% that of liver (42,43). Other tissues exhibit activities <4% that of liver (42). As discussed above, the enzyme provides a means of shuttling the nitrogen of certain amino acids, through linked transaminase reactions, toward ammonia for urea synthesis in liver. The role of glutamate dehydrogenase in brain is less clear. Tracer studies suggest that glutamate dehydrogenase is not quantitatively important for the removal of ammonia in normal (11) and moderately hyperammonemic rat brain (44), but may become so at very high (toxic) levels of ammonia (45). In brain, glutamate dehydrogenase is present in both astrocytes and neurons. Indeed, it has been proposed that astrocytic glutamate dehydrogenase plays a special role in metabolizing glutamate (46) and perhaps in neutralizing released neurotransmitter glutamate (47). However, other evidence suggests that aspartate aminotransferase may be more important than glutamate dehydrogenase in metabolizing astrocytic glutamate (48,49). Glutamate dehydrogenase is so active in liver that the components of the reaction appear to be at thermodynamic equilibrium (42); however, this cannot be demonstrated in brain, possibly because of compartmentation of glutamate in different cell types (50). It seems probable that because of the inherently high activity of glutamate dehydrogense in the brain, the reaction will generate or degrade glutamate according to the metabolic dictates of the cell. The ability of the brain cell to degrade glutamate by glutamate dehydrogenase will depend to some extent on the capacity to remove ammonia and pull the reaction in the direction of glutamate removal.

AMP deaminase (equ. 4) is especially abundant in muscle. As ATP is hydrolyzed to ADP, ATP can be formed by the adenylate kinase (myokinase) reaction (2ADP <---> ATP + AMP). The reaction is freely reversible but is pulled in the direction of ATP formation by the action of AMP deaminase. AMP is regenerated from IMP by the action of adenylosuccinate synthetase (equ. 5) and adenylosuccinate lyase (equ. 6).

AMP + H_2O	->	IMP + NH_3	(4)
IMP + aspartate + GTP	->	adenylosuccinate + GTP + Pi	(5)
Adenylosuccinate	->	AMP + fumarate	(6)
Net: Aspartate + GTP + H_2O	->	fumarate + GDP + Pi + NH_3	(7)

The three enzymes (Equs. 4–6) are collectively known as the purine nucleotide cycle (PNC) (51). The cycle appears to be a major source of ammonia in muscle (52), and brain (53), an alternative source of ammonia in kidney (54) (but only at high aspartate concentrations (55)) and of low importance in liver (39). In the rat, adenylate kinase activity is greatest in skeletal muscle, followed by heart and brain, with smaller amounts in kidney and liver (56). Provision of additional energy to exercising muscle or to firing neurons from the myokinase/AMP deaminase reaction is probably of limited duration. Nevertheless, the PNC is physiologically very useful. Lowenstein (51) has provided strong evidence that the pathway is important: 1) in liberating ammonia from amino acids, 2) in adjusting the level of TCA intermediates, 3) as a source of carbon for energy production from certain amino acids, 4) in regulating ATP, ADP and AMP levels, 5) in control of glycolysis. The role of the PNC in ammonia production from some amino acids is best exemplified by the following series of equations:

L–Amino acid + α–ketoglutarate	⇌	α–keto acid + L–glutamate	(1)
L–Glutamate + oxaloacetate	⇌	α–ketoglutarate + L–aspartate	(8)
L–Aspartate + GTP + H_2O	->	fumarate + GDP + Pi + NH_3	(7)
Fumarate + H_2O	⇌	malate	(9)
Malate + NAD^+	⇌	oxaloacetate + NADH + H^+	(10)
Net: L–Amino acid + NAD^+ + GTP + $2H_2O$	->	α–keto acid + NADH + GDP + Pi + NH_4^+	(11)

In this case, the net reaction for the formation of ammonia from an amino acid (equ. 11) is more favorable than the transdeamination reaction (equ. 3) since it is coupled to the hydrolysis of GTP. Note that if equ. 1 is omitted, the net reaction is a deamidation of glutamate (equ. 12) that is energetically more favorable than the glutamate dehydrogenase reaction.

L–Glutamate + NAD^+ + GTP + $2H_2O$ -> α–ketoglutarate + NADH + GDP + Pi + NH_4^+ (12)

The major pathways involving ammonia metabolism are shown in Fig 2.

Other sources of ammonia will be discussed briefly. The kidney extracts glycine from the blood and converts it to ammonia and serine (equ. 15) via the combined action of the glycine cleavage system (equ. 13) and serine transhydoxymethyltransferase (equ. 14) (57). The serine is released to the blood to be metabolized by the liver.

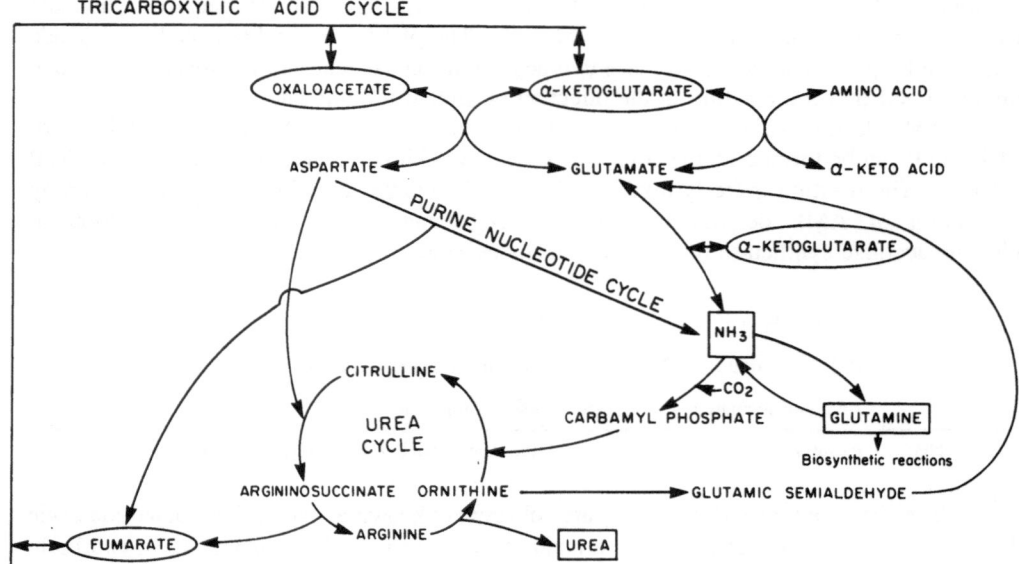

Figure 2. Scheme linking the major ammonia producing and consuming reactions of the body. TCA cycle intermediates are circled. The major fate of the nitrogen of those amino acids that are degraded in a first step catalyzed by transamination (e.g., alanine, tyrosine, branched–chain amino acids) is indicated by boxed products. Note that the scheme is a composite because while the TCA cycle is present throughout the body (with the notable exception of red blood cells), the complete urea cycle is almost totally confined to the liver and the PNC is important in some extrahepatic tissues. From 15 with permission.

$$\text{Glycine} + NAD^+ + H_4 \text{ folate} \quad \rightarrow \quad NH_4^+ + CO_2 + N^5,N^{10}\text{–}CH_2\text{–}H_4 \text{ folate} + NADH \quad (13)$$

$$\text{Glycine} + N^5,N^{10}\text{–}CH_2\text{–}H_4 \text{ folate} \quad \rightleftarrows \quad \text{L–serine} + H_4 \text{ folate} \quad (14)$$

$$\text{Net:} \quad 2 \text{ Glycine} + NAD^+ \quad \rightarrow \quad \text{L–serine} + NH_4^+ + CO_2 + NADH \quad (15)$$

It has been calculated that in the normal, adult human male about 63, 24, and 13% of the ammonia produced in the kidney is generated from glutamine (amide), glycine and glutamate/glutamine (amine), respectively (58). Of this ammonia, one half enters the blood stream and the other half enters the urine (58). The importance of glutamine (amide) as a source of renal ammonia is obviously well established. However, there are two possible routes to ammonia formation from glutamine (amide). The first, phosphate–activated glutaminase (PAG) has already been mentioned. The second is the glutamine transaminase –ω–amidase pathway (glutaminase II) (equs. 16 –18).

$$\text{L–Glutamine} + \alpha\text{–keto acid} \quad \rightleftarrows \quad \alpha\text{–ketoglutaramate} + \text{L–amino acid} \quad (16)$$

$$\alpha\text{–Ketoglutaramate} + H_2O \quad \rightarrow \quad \alpha\text{–ketoglutarate} + NH_3 \quad (17)$$

$$\text{Net: L–Glutamine} + \alpha\text{–keto acid} + H_2O \rightarrow \quad \alpha\text{–ketoglutarate} + \text{L–amino acid} + NH_3 \quad (18)$$

Like PAG, glutamine transaminase occurs in a liver (or L)– and a kidney (or K)–form. In rats, the L form is largely restricted to liver but the K form is widespread (59, 60). [Glutamine transaminase K has attracted considerable attention recently among pharmacologists because the enzyme is now known to be a major cysteine–S–conjugate β–lyase of the kidney, and is responsible for converting certain halogenated cysteine conjugates to nephrotoxic compounds (61 and references quoted therein)]. ω–Amidase is widespread in rat tissues, roughly paralleling glutamine transaminase activity (60), suggesting perhaps a coordinated gene expression. Most investigators (e.g., 37,62) assume that the glutaminase II pathway to ammonia formation in the kidney is unimportant. However, it is difficult to distinguish the PAG from the glutaminase II pathway by tracer studies. Calculations show that under optimal conditions of assay the specific activity of rat kidney glutamine transaminase K is about 24% that of PAG (61). The glutamine transaminase of rat kidney is restricted to the S1, S2, and S3 regions of the nephron (63, Cooper A.J.L., and Endou H, unpublished results). In the rat brain, the glutamine transaminase K – ω –amidase pathway is of high activity in the choroid plexus (unpublished data). The choroid plexus in brain and S1, S2, and S3 regions in kidney are involved in amino acid recycling. It is possible that glutamine transaminase K is intimately involved in this process. While it is probable that most of the ammonia derived from glutamine (amide) in brain and kidney is generated from the PAG reaction, it is possible that the glutamine transaminase – ω –amidase pathway is important locally for the generation of ammonia in these organs.

5. Interorgan Flux of Nitrogen

The major metabolic fates of the common protein amino acids are shown in figs. 3 and 4. The central role of ammonia in these processes is highlighted. For simplicity only the nitrogen is traced. An excellent quantitative analysis of amino acid oxidation and related gluconeogenesis has recently been published (58). The fate of the amino carbon is discussed in great detail in this review, so only a few points concerning the fate of amino acid carbon will be made here. Most of the amino acids that arise from digestion of dietary protein and that are not required for protein synthesis and other biosynthetic processes are oxidized in the liver. Oxidation of amino acids provides the liver with 50% of its energy needs (58). However, oxidation is not complete and amino acid carbon is mostly incorporated into glucose. The advantage of producing glucose in the liver by these means is that extrahepatic organs can make use indirectly of two–thirds of the energy available from protein fuels obtained in the diet and yet be spared the task of synthesizing amino acid–catabolizing enzymes (58). Unlike most amino acids, the catabolism of the branched–chain amino acids (BCAAs) does not occur (or is not initiated) in the liver. These amino acids are taken up by the skeletal muscles and transaminated. In many animals the corresponding α–keto acids are released to the circulation to be oxidized in the liver; however, in humans the bulk of these α–keto acids is oxidized in situ in the muscle (64). The positioning of the machinery for the catabolism of the BCAAs in the musculature (and to a lesser extent in brain) may be to provide a "safety net". If the BCAAs were also oxidized in the liver, the total burden of oxidizing all excess amino acids would severely limit the amount of O_2 available for other oxidative processes in the liver (58). Moreover, there is evidence that our Paleolithic ancestors obtained about a third of their dietary calories from protein (58,65) as compared with 15% in the modern American diet. Jungas et al point out that such a diet could only be tolerated if the BCAAs were oxidized outside the liver (58). In the following section I discuss nitrogen homeostasis separately for several major tissues and point out how these processes are integrated in the whole body. A similar approach was used in a previous review (66).

5.1. Muscle. Nitrogen intake is mostly in the form of BCAAs. The major fate of the nitrogen is incorporation into alanine (through linked aminotransferase (AT) reactions) and into glutamine (amide) via ammonia. Theoretically, muscle ammonia can be generated from the BCAAs through the transdeamination pathway (equ. 3) or via the PNC (equ. 11). AMP is an important source of ammonia in muscle (51 and refs. cited therein) and much evidence

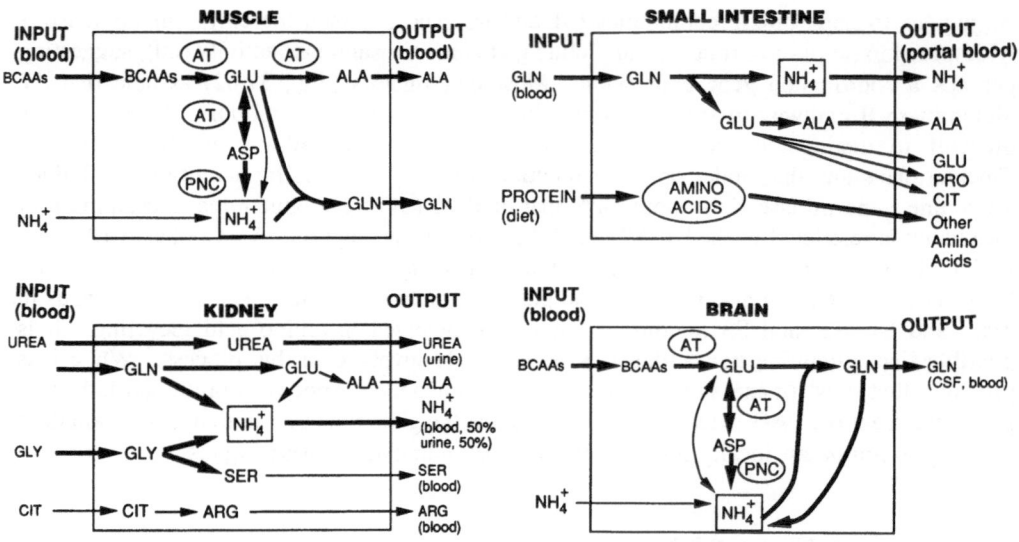

Figure 3. Nitrogen metabolism in some extrahepatic tissues. The most prominent pathways are indicated by a thickened arrow. For a detailed treatment of the metabolism of amino acid carbon see ref. 58. AT, aminotransferases; PNC, purine nucleotide cycle.

suggests that the PNC is an important route to the production of ammonia in resting and exercising muscle (52). Muscle is generally thought to contain little glutamate dehydrogenase and associated transdeamination activities. [However, a recent report suggests that human muscle contains enough glutamate dehydrogenase to fuel a transdeamination–type reaction (67), and the importance of the PNC in the generation of ammonia in exercising human muscle has been questioned (68).] Felig (69) has proposed that alanine lost from the musculature is converted to glucose in the liver and that glucose returning from the liver acts as a source of pyruvate to be transaminated to alanine – the "glucose – alanine cycle". However, it is likely that other sources in addition to glucose can provide carbon lost as alanine (15,58). Moreover, glutamine output far exceeds that of alanine in human muscle (64).

Lockwood et al. showed that in normal volunteers 50% of injected [^{13}N]ammonia is removed by the musculature (17). This finding suggests that muscle is able to "fix" appreciable amounts of blood–borne ammonia. Indeed, a small net uptake of ammonia (11%) has been demonstrated across normal human leg muscle (70). A net uptake of ammonia across forearm muscle has also been noted in patients with liver disease but not in controls (71). Other workers have been unable to demonstrate a net uptake of ammonia across the rat hindquarters (72). Evidently, muscle consumes (as glutamine) and generates (probably via the PNC) ammonia, such that on balance, at least in the human, there is a small net uptake of ammonia in resting muscle. However, although glutamine synthetase of resting muscle is a major "sink" for the removal of endogenous ammonia generated in extrahepatic tissues, activity is limited and blood ammonia levels will rise during sustained exercise (e.g. 70) or during liver disease (13,66,72).

Glutamine transport in muscle is mediated by a Na$^+$ –dependent carrier that is shared by asparagine and histidine (73). The carrier has been designated N^m and is different from the glutamine or (N) carrier of the liver. Hundal et al. suggest that the presence of this carrier can explain the large ratio of muscle to plasma glutamine and suggest that whole–body glutamine metabolism may be modulated by muscle glutamine transport (73). Glutamine efflux from muscles is enhanced in a number of severe metabolic insults such as starvation, acidosis (see below), trauma, sepsis and burns (74).

The rat heart possesses the enzymatic machinery to synthesize and catabolize glutamine (75 and references cited therein). Whether the heart is a net exporter or consumer of glutamine remains to be established (75). Interestingly, however, the human heart has been shown to release alanine (76).

5.2. Brain. The brain contains two metabolically distinct compartments. Ammonia is converted to glutamine preferentially in a small pool of rapidly turning over glutamate that is distinct from a larger, more slowly turning over glutamate pool. The brain also contains two distinct TCA cycles. The small and large pools are now known to be largely represented by astrocyes and neurons, respectively (for reviews see refs. 13 and 29). As noted above, blood–borne ammonia enters the brain mostly by diffusion to be incorporated into glutamine (amide) within the astrocytes (11). Because astrocytes surround the capillaries one would predict a positive A–V difference for ammonia across brain. Such a difference has not been observed for normal rat brain, but has been demonstrated for hyperammonemic rat brain (77) and for the brains of a number of species (13) including man (78).

The blood–brain barrier (BBB) possesses at least three amino acid transporters: a) a high affinity, high capacity carrier for large neutral amino acids, b) a high affinity, medium capacity carrier for basic amino acids, and c) a high affinity, low capacity carrier for acidic amino acids (79,80). The first two carriers supply the brain with essential amino acids. The role of the third carrier is less clear as the brain can synthesize glutamate and aspartate in large amounts. Obviously, the brain must take up basic and large, neutral amino acids to meet the demands of protein synthesis and, in some cases, of neurotransmitter synthesis. However, it appears that, quantitatively, of the amino acids taken up by the brain the BCAAs are the most important (13). The BCAA nitrogen is transferred to glutamate in the astrocytes. To offset ingress of nitrogen as ammonia and amino acids, the brain exports glutamine to the CSF and to the blood (13,77,81).

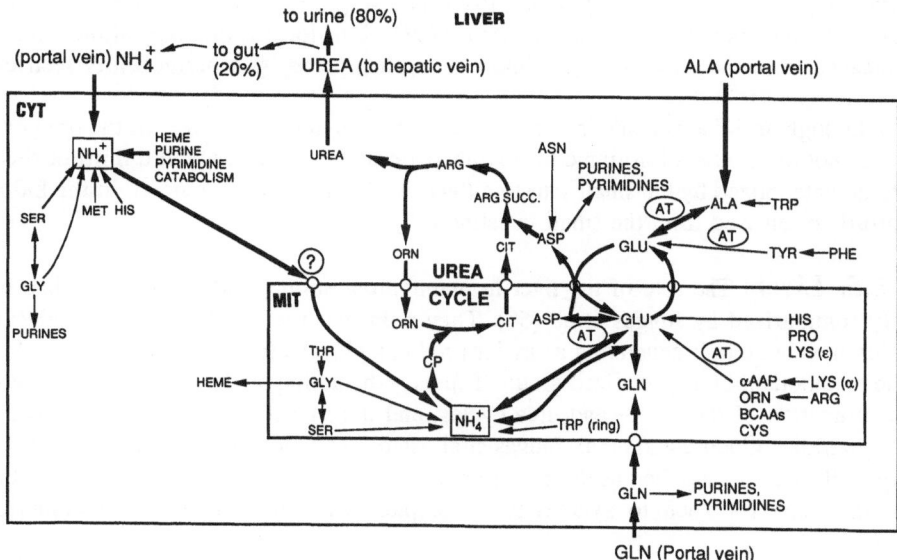

Figure 4. Nitrogen metabolism in liver. Excess nitrogen from the breakdown of the twenty commonly occuring protein amino acids is incorporated into urea via the pathways shown. Note that some amino acid nitrogen is incorporated into glutamine, alanine and serine in some extrahepatic tissues (Fig.3). Also shown is the importance of the liver in heme, purine and pyrimide homeostasis. Note the important link between urea– and TCA cycles. The small circles indicate mitochondrial transport systems. The question mark indicates that there is some uncertainty as to whether there is a transporter for ammonia. CYT, cytosol; MIT, mitochondrion.

In addition to skeletal muscle and brain other organs such as lung (29,74,82), skin (83,84), and adipose tissue (83,84) contribute glutamine to the circulation.

5.3. Gut. Amino acids arising from the digestion of proteins in the diet are released to the circulation via the portal vein to be used as needed; as noted above, most surplus amino acids, with the notable exception of the BCAAs, are degraded in the liver. Glutamine released from the muscles (and to a lesser extent from other organs) is taken up by the gut (e.g. 74,85–87) wherein it is a major energy source (22,88,89). Glutamine uptake in the small intestine is mediated by the Na^+-dependent A system (74 and refs. quoted therein). Metabolism of glutamine in the small intestine results in the release of large amounts of alanine (a gluconeogenic precursor in liver) and ammonia, some citrulline and smaller amounts of proline and glutamate (89). The oxidation of glutamine may be involved in the regulation of water and Na^+/K^+ transport in the small intestine (90).

5.4. Kidney. Circulating glutamine is taken up by the kidney to be converted into ammonium. Secretion of ammonium ions to the urine plays an important role in acid–base balance. Indeed, it is well known that during acidosis the kidney extracts more glutamine from the blood and produces more ammonium ions (e.g., 87,91). During acidosis the blood level of glutamine remains constant so that in order to maintain blood glutamine levels some other organ(s) must increase glutamine output. In rats, acidosis stimulates increased output of glutamine from muscle, and liver switches from a net consumer of glutamine to a net producer (87). In dogs, however, muscle glutamine output does not increase in acidosis and the liver provides the increased glutamine load (91). As mentioned above, the kidney is known to take up glycine and contribute serine to the blood. In addition, the rat kidney may take up small amounts of BCAAs and liberate alanine to the circulation (92). According to Jungas et al. (58) net conversion of glutamine to serine ensures equiproduction of ammonium and bicarbonate by the kidney: since half of this ammonium is excreted, the kidney will export a small excess of bicarbonate to the blood. Of interest is the recent finding that (contrary to previous claims) pig renal PAG has a predominantly functional external localization in the inner mitochondrial membrane (93). This localization may allow ammonia to be released from the mitochondria without being hampered by a permeability restriction (93).

Although arginine is made in considerable amounts during turnover of the urea cycle, the liver is not the source of arginine for extrahepatic tissues. Also, about a third of dietary arginine is metabolized by the small intestine (94). Arginine is made de novo in the kidneys from citrulline released from the small intestine (95).

5.5. Liver. The role of the liver in maintaining amino acid homeostasis has been elegantly summarized by Jungas et al. (58). They point out that the liver can be viewed as a complex metabolic unit generating as major products glucose (or glycogen), urea and CO_2 from the metabolism of a specialized group of fuels – the amino acids. The process requires close coordination of the TCA– and urea cycles, and is self sufficient despite the need for ATP to synthesize glucose and urea. Jungas et al. state: "The most useful outcome is that the majority of the energy supplied as dietary protein is made available to all tissues of the body without the need for each to synthesize a complex array of amino acid–metabolizing enzymes".

We are now in a position to outline the fate of the nitrogen of all twenty of the common amino acids arising from protein digestion and to summarize the role of the liver in this process. [Some of the conclusions are based on the work of Jungas et al. (58), but others are based on some of the author's findings or are from additional sources]. ALA: Alanine is extensively oxidized by liver. Because of net output by muscle and gut the amount of alanine consumed by the liver is several-fold greater than that obtained in the diet. ASP, GLU:

Dietary glutamate and aspartate are mostly metabolized by the small intestine (94). In most tissues, glutamate and aspartate are readily obtained from transamination of α–ketoglutarate and oxaloacetate, respectively, so that there is no need to appreciably shuttle these amino acids among different organs. Nevertheless, there is some release of glutamate from the gut and uptake of this amino acid by the perivenous cells of the liver (39,96). BCAAs: These are metabolized by skeletal muscle although smaller amounts may be metabolized by brain and kidney. GLY: About 40% is metabolized by the kidney and 60% by the liver (58). SER: Serine is extensively metabolized by the liver. Because of serine production by the kidney the amount of this amino acid consumed by the liver is about twice that obtained in the diet (58). ARG: Dietary arginine escaping metabolism in the small intestine and arginine produced by the kidney are metabolized in the liver. GLN: This amino acid is discussed separately, below. Net degradation of all other amino acids (ASN, CYS, HIS, LYS, MET, PHE, PRO, THR, TRP, TYR) is largely carried out in the liver. The major fate of the nitrogen of each of these amino acids is summarized in Fig. 4.

As noted above, glutamine is taken up by the periportal hepatocyes and released by the perivenous hepatocytes such that under normal conditions there is a small uptake of glutamine. Jungas et al. (58) estimate that only 2% of all net amino acid nitrogen metabolized by the liver is in the form of glutamine (amide). The importance of glutamine as a precursor of urea nitrogen has been debated (see refs. quoted in 39). However, our studies with L–[amide $-^{13}$N]glutamine unequivocally show that the normal, rat liver does indeed convert glutamine (amide) nitrogen to urea in vivo. We estimate that glutamine (amide) contributes modestly (7 –12%) to urea nitrogen homeostasis in the normal, fed rat (39). The activation of liver PAG by ammonia (for which there is an absolute requirement) is complex. In the isolated perfused liver, half–maximal and maximal activation of glutaminase is obtained at 0.2 μM and 0.6 μM NH_4^+ in the perfusate (97,98). Since these concentrations occur in the portal vein, the portal ammonia concentration may be an important regulator of hepatic glutamine breakdown (99). In other words, ammonia released from the gut is an important signal for glutamine degradation in the liver (100). Liver PAG is inhibited by low pH (101,102). This inhibition is due to a decrease in ammonia stimulation and to a diminished transport of glutamine across the plasma and mitochondrial membranes (101–102). Häussinger has pointed out that the unique properties of liver glutaminase allows this enzyme to function as a pH–modulating ammonia amplifying system inside the mitochondrion (99). This system allows for effective urea synthesis at low portal vein ammonia concentrations. According to Häussinger, liver PAG provides additional ammonia to the mitochondria, helping to offset the relatively high Km exhibited by carbamyl phosphate synthetase I for ammonia. Szweda and Atkinson (102) showed that liver PAG activity responds strongly to physiological concentrations of citrate at physiological concentrations of NH_4^+, Pi and glutamine. The authors suggested that the findings are consistent with the view that liver PAG has a regulatory role in ureagenesis.

Complete oxidation of amino acids results in roughly equivalent formation of CO_2, HCO_3^- and NH_4^+. In the adult human male 1.0 – 1.4 moles of HCO_3^- are produced daily from the metabolism of amino acids (58,103,104). Atkinson and Bourke (103, 104) have argued that because the pKa of ammonia is 9.2, ammonium ions cannot effectively buffer against bicarbonate; therefore, the urea cycle is necessary to prevent severe alkalosis and plays a prominent role in the regulation of acid – base balance. Most biochemistry and physiology texts, until recently, have discussed the urea cycle solely in terms of its role in removing waste nitrogen. However, because both HCO_3^- and NH_4^+ are converted to neutral urea the urea cycle must perforce be involved to some extent in acid–base homeostasis. The question then becomes "Is this a major function of the urea cycle as Atkinson has suggested or an incidental role?" Häussinger and colleagues (105, 106) have strongly endorsed the idea that the urea cycle is involved in whole body acid–base balance whereas others have been strongly critical (107). If NH_4^+ and HCO_3^- are produced in equal amounts and converted to urea the sytemic pH is not affected. However, in metabolic acidosis that is not the result of renal insufficiency, HCO_3^- is retained and NH_4^+ is excreted in the urine (33). According to Häussinger and colleagues, by regulating the flux of ammonia into urea or glutamine the liver becomes an important organ for the maintenance of HCO_3^-. During acidosis, NH_4^+ excretion is increased, HCO_3^- is spared and the systemic pH rises. The acidosis–induced increase in NH_4^+ excretion is brought about first by increased incorporation into glutamine in the liver which upon release to the circulation is taken up by the kidney and converted therein to glutamate and NH_4^+ by kidney glutaminase. Glutamine acts as a nontoxic transport from of ammonia between liver and kidney. The true substrate of carbamyl phosphate synthetase I is NH_3 and not NH_4^+ (108). Urea synthesis increases with NH_3 concentration in the plasma (presumably $[NH_3]$ in liver cells is proportional to $[NH_3]$ in plasma) and not necessarily with pH (109,110). In a recent review, Meijer et al. (33) argue that urea synthesis is not an exclusive mechanism for HCO_3^- detoxification as proposed by Atkinson but, in addition, still serves as a major mechanism for NH_4^+ detoxification in agreement with the traditional view.

Nevertheless, it is the opinion of Meijer et al. (33) that the importance of the urea cycle in pH homeostasis (inhibition leads to retention of HCO_3^-) has been established beyond reasonable doubt. The authors point out that, in contrast to the traditional view, excretion of NH_4^+ cannot be considered as excretion of acid, since during amino acid catabolism NH_4^+ and not NH_3 is formed. They further argue that control of systemic pH occurs by retention of HCO_3^- not used in urea synthesis. The function of the kidney is therefore to remove surplus NH_4^+ not used in urea synthesis. This does not mean that the kidney has a diminished role relative to the traditional view but rather the kidney is important in NH_4^+ homeostasis more than in pH homeostasis (33). Perhaps, it is best, in fact, to regard pH homeostasis as being regulated by a complex interplay of several organs: liver (urea cycle), kidney (NH_4^+ removal), lung (CO_2 removal) and possibly muscle (33).

6. Conclusions

Ammonia is generated in extrahepatic tissues (Fig. 3) and liver (Fig. 4) from the breakdown of amino acids and other nitrogenous substances. Ammonia is also a key building block in the synthesis of a number of important biomolecules. Despite its central importance in nitrogen homeostasis ammonia is toxic at elevated concentrations, especially to the central nervous system (13) and its concentration must be kept low. The body has adapted by utilizing various interorgan "nitrogen shuttles" that are integral to the regulation of nitrogen homeostasis. Although the general outlines of amino acid/ammonia metabolism are now well known much work remains to be done to further define these pathways in nomal and various disease states.

References

1. Moat A., and Foster J.W., 1989 Microbial Physiology, 2nd Edition. John Wiley and Sons, New York, pp. 251 – 283.
2. Tannenbaum S.R., Fett D., Young V.R., Land P.D., and Bruce W.R. 1978 Nitrite and nitrate are formed by endogenous synthesis in the human intestine. Science 200:487 – 1489.
3. Parks N.J., Krohn K.R., Mathis C.A., Chasko J.H., Greiger K.R., Gregor M.E., and Peek N.F. 1981 Nitrogen–13–labeled nitrite and nitrate: Distribution and metabolism after intratracheal administration. Science 212:58 –61.
4. Thayer J.R., Chasko J.H., Swartz L.A., and Parks N.J., 1982 Gut reactions of radioactive nitrite after intratracheal administration in mice. Science 217:151 – 153.
5. Xie Q., Cho H.J., Calaycay J., Mumford R.A., Swiderek, K.M., Lee T.D., Ding A., Troso T., and Nathan C., 1992 Cloning and characterization of inducible nitric oxide synthase from mouse macrophages. Science 256:225 – 228.

6. Sillún L.G. 1964 Stability Constants of Metal–Ion Complexes. 1. Inorganic ligands. Chemical Society, London, p.150. Special Publ. 17.

7. Bromberg P.A., Robin E.D., and Forkner C.E. Jr. 1960 The existence of ammonia in blood in vivo and the significance of the NH_4^+–NH_3 system. J. Clin. Invest. **39:**332 – 341.

8. Dutton R.E., and Berkman R.A., 1978 Ammonia and the regulation of ventilation. In: Regulation of Ventilation and Gas Exchange, edited by D.G. Davis and C.D. Barnes. Academic Press, New York, pp. 69 – 91.

9. Jacquez J.A., Poppel J.W., and Jeltsch R., 1959 Solubility of ammonia in human plasma. J. Appl. Physiol. **14:**255 – 258.

10. Klocke R.A., Anderson K.K., Rotman H.H., and Forster R.E, 1972 Permeability of human erythrocytes to ammonia and weak acids. Am J. Physiol. **222:**1004 – 1013.

11. Cooper A.J.L., McDonald J.M., Gelbard A.S., Gledhill R.F., and Duffy T.E. 1979 The metabolic fate of 13N–labeled ammonia in rat brain. J. Biol. Chem. **254:**4982 –4992.

12. Raichle M.E., and Larson K.B. 1981 The significance of the NH_3 – NH_4^+ equilibrium on the passage of 13N–ammonia from blood to brain. A new regional residue detection method. Circ. Res. **48:**913 – 937.

13. Cooper A.J.L., and Plum F. 1987 Biochemistry and physiology of brain ammonia. Physiol. Rev. 67: 440 – 519.

14. Braunstein A.E., 1957 Les voies principal de l'assimilation de l'azote chez les animaux. Adv. Enzymol. **19:**335 – 389.

15. Cooper A.J.L., 1988 L–Glutamate 2–oxoglutarate aminotransferases. In: Glutamine and Glutamate in Mammals. Vol 1., edited by E. Kvamme, CRC Press, Boca Raton, Fl., pp.123 – 152.

16. Pamiljans V., Krishnaswamy P.R., Dumville G., and Meister A., 1962 Studies on the mechanism of glutamine synthesis: Isolation and properties of the enzyme from sheep brain. Biochemistry **1:**153 – 158.

17. Lockwood A.H., McDonald J.M., Reiman R.E., Gelbard A.S., Laughlin J.S., Duffy T.E., and Plum F., 1979 The dynamics of ammonia production in man. Effects of liver disease and hyperammonemia. J. Clin. Invest. **63:**449 – 460.

18. Rowe W.B. 1985 Glutamine synthetase from muscle. Methods Enzymol. **113:**199 – 212.

19. Magnuson S.R., and Young A.P. 1988 Murine glutamine synthetase: Cloning, developmental regulation, and glucocorticoid inducibility. Develop. Biol. **130:**536 – 542.

20. Herbert J.D., Coulson R.A., and Hernandez T. 1966 Free amino acids in the caiman and rat. Comp. Biochem. Physiol. **17:**583 – 598.

21. Record C.O., Buxton B., Chase R.A., Curzon G., Murray–Lyon I.M., and Williams R., 1976 Plasma and brain amino acids in fulminant hepatic failure and their relationship to hepatic encephalopathy. Eur. J. Clin. Invest. **6:**87 – 394.

22. Windmueller H.G., and Spaeth A.E. 1980 Respiratory fuels and nitrogen metabolism in vivo in small intestine of fed rats. J. Biol. Chem. **255:**107 – 112.

23. Biltz R.M., Letteri J.M., Pellegrino E.D., Palekar A., and Pinkus L.M. 1983 Glutamine metabolism in bone. Min Electrolyte Metab. **9:**25 – 131.

24. Zielke H.R., Ozand P.T., Tildon J.T., Sevdalian D.A.,and Cornblath M.,(1980 Reciprical regulation of glucose and glutamine utilization by cultured human diploid fibroblasts. J. Cell Physiol. 95: 41 – 48.

25. Pitts R.F.,(1975 Production of CO_2 by the intact functioning kidney of the dog. In: The Medical Clinics of North America. Vol 59:3. Symposium on Renal Metabolism., edited by S. Baruch, W.B. Saunders Co., Philadelphia, pp. 507 – 518.

26. Norenberg M.D., and Martinez–Hernandez A.,(1979 Fine structural localization of glutamine synthetase in astrocytes of rat brain. Brain Res. 161: 303 – 310.

27. Benjamin A.M., and Quastel J.H., 1975 Metabolism of amino acids and ammonia in rat brain cortex slices in vitro: A possible role of ammonia in brain function. J. Neurochem. **25:**197 – 206.

28. Hertz L.,1979. Functional interaction between neurons and astrocytes. 1. Turnover and metabolism of putative amino acid neurotransmitters. Progr. Neurobiol. **13:**277 – 323.

29. Cooper A.J.L. Glutamine synthetase. In: Glutamine and Glutamate in Mammals. Vol. 1., edited by E. Kvamme, CRC Press, Boca Raton, Fl, pp.7 – 31.

30. Cooper A.J.L., Vergara F., and Duffy T.E., 1983 Cerebral glutamine synthetase. in: Glutamine, Glutamate and GABA in the Central nervous System, edited by L. Hertz, E. Kvamme, E.G.

McGeer, and A. Schousboe, Alan R. Liss, Inc., New York, pp.77 – 93.

31. Tate S.S., and Meister A., 1973 Glutamine synthetases of mammalian liver and brain. In: The Enzymes of Glutamine metabolism, edited by S. Prusiner and E.R. Stadtman, Academic Press, New York, pp. 77 – 127.

32. Hőussinger D., 1983 Hepatocyte heterogeneity in glutamine and ammonia metabolism and the role of an intracellular glutamine cycle during ureogenesis in perfused rat liver. Eur. J. Biochem. **133**:269 – 275.

33. Meijer A.J., Lamers W.H.,, and Chameleau R.A.F.M.,(1990 Nitrogen metabolism and ornithine cycle function. Physiol. Rev. **70**:701 – 748.

34. Jungermann K., and Katz N., 1989 Functional specialization of different hepatocyte populations. Physiol. Rev. **69**:708 – 764.

35. Cooper A.J.L., Nieves E., Coleman A.E., Filc–DeRicco S., and Gelbard A.S., 1987 Short–term metabolic fate of [^{13}N]ammonia in rat liver in vivo. J. Biol. Chem. **262**:1073 – 1080.

36. Smith E.M., and Watford M.,(1990 Molecular cloning of a cDNA for rat hepatic glutaminase. Sequence similarity to kidney glutaminase. J. Biol. Chem. **265**:10631 – 10636.

37. Tannen R.L., and Sahai A., 1990 Biochemical pathways and modulators of renal ammoniagenesis. Min. Electolyte Metabol. **16**:249 –258.

38. Banner C., Hwang J–J., Shapiro R.A., Wenthold R.J., Nakatani Y., Lampel K.A., Thomas J.W., Huie D., and Curthoys N.P., 1989 Isolaiton of a cDNA for rat brain glutaminase. Mol. Brain Res. **3**:247 – 254.

39. Watford M., and Smith E.M., 1990 Distribution of hepatic glutaminase acivity and mRNA in perivenous and periportal hepatocytes. Biochem. J. **267**:265 – 267.

40. Cooper A.J.L., Nieves E., Rosenspire K.C., Filc–DeRicco S., Gelbard A.S., and Brusilow S.W., 1988 Short–term metabolic fate of ^{13}N–labeled glutamate, alanine and glutamine (amide) in rat liver. J. Biol. Chem. **263**:12268 – 12273.

41. Meijer A.J., 1985 Channeling of ammonia from glutamine to carbamoyl–phosphate synthetase in liver mitochondria. FEBS lett. **191**:249 – 251.

42. Williamson D.H., Lund P., and Krebs H.A., 1967 The redox state of free nicotinamide–adenine dinucleotide in the cytoplasm and mitochondria of rat liver. Biochem. J. **103**:514 – 527.

43. Chee P.Y., Dahl J.L., and Fahien L.A., 1979 The purification and properties of rat brain glutamate dehydrogenase. J. Neurochem. **33**:53 – 60.

44. Cooper A.J.L., Mora S.N., Cruz N.F., and Gelbard A.S., 1985 Cerebral ammonia metabolism in hyperammonemic rats. J.Neurochem. **44**:1716 – 1723.

45. Berl S., Takagaki G., Clarke D.D., and Waelsch H., 1962 Metabolic compartmentation in vivo: Ammonia and glutamic acid metabolism in brain and liver. J. Biol. Chem. **237**:2562 – 2569.

46. Yu A.C.H., Schousboe A., and Hertz L., 1983 Metabolic fate of ^{14}C–labeled glutamate in astrocytes in primary culture. J. Neurochem. **39**:954 – 960.

47. Aoki C., Milner T.A., Berger S.B., Sheu K.–F.R., Blass J.P., and Pickel V.M., 1987 Glial glutamate dehydrogenase: Ultrastructural localization and regional distribution in relation to the mitochondrial enzyme, cytochrome oxidase. J. Neurosci. Res. **18**:305 – 318.

48. Lai J.C.K., Murthy Ch.R.K., Cooper A.J.L., Hertz E., and Hertz L., 1989 Differential effects of ammonia and 6–methylene–D,L–aspartate on metabolism of glutamate and related amino acids by astrocytes and neurons in primary culture. Neurochem. Res. **14**:377 – 389.

49. Farinelli S.E., and Nicklas W.J., 1992 Glutamate metabolism in rat cortical astrocyte cultures. J. Neurochem. **58**:1905 – 1915.

50. Howse D.C., and Duffy T.E., 1975 Control of the redox state of the pyridine nucleotides in the rat cerebral cortex. Effect of electroshock–induced seizures. J. Neurochem. **24**:935 –940.

51. Lowenstein J.M., 1972 Ammonia production in muscle and other tissues: The purine nucleotide cycle. Physiol. Rev. **52**:382 – 414.

52. Lowenstein J.M., 1990 The purine nucleotide cycle revisited. Int. J. Sports Med. 11 Suppl. **2**:S37 – S47.

53. Schultz V., and Lowenstein J.M., 1978 The purine nucleotide cycle. Studies of ammonia production and interconversions of adenine and hypoxanthine nucleotides and nucleosides by rat brain in vivo. J. Biol. Chem. **253**:1938 – 1943.

54. Bogusky R.T., and Aoki T.T., 1983 Early events in the initiation of ammonia formation in the kidney. J. Biol. Chem. **258**:2795 – 2801.

55. Tamura K., and Endou H., 1988 Contribution of purine nucleotide cycle to intranephron

ammoniagenesis in rats. Am J. Physiol. **255**:F1122 – F 1127.

56. Oliver I.T., 1955 A spectrophotometric method for the determination of creatine phosphokinase and myokinase. Biochem. J. **61**:116 – 122.

57. Lowry M., Hall D.E., and Brosnan J.T., 1986. Serine synthesis in rat kidney: Studies with perfused kidney and cortical tubules. Am. J. Physiol. **250**:F649 – F658.

58. Jungas R.L., Halperin M.L., and Brosnan J.T., 1992 Quantitative analysis of amino acid oxidation and related gluconeogenesis in humans. Physiol. Rev. **72**:419 – 448.

59. Cooper A.J.L., and Meister A., 1981 Comparative studies of glutamine transaminases from rat tissues. Comp. Biochem. Physiol. **69B**:137 – 145.

60. Cooper A.J.L. Glutamine aminotransferases and ω–amidases. In: Glutamine and Glutamate in Mammals. Vol 1., edited by E. Kvamme, CRC Press, Boca Raton, FL., pp. 33 – 52.

61. Cooper A.J.L., and Anders M.W., 1990 Glutamine transaminase K and cysteine conjugate 6–lyase. Ann. N. Y. Acad. Sci. **585**:118 – 127.

62. Nissim I., Wehrli S., States B., Nissim I., and Yudkoff M., 1991 Analysis and physiological implications of renal 2–oxoglutaramate metabolism. Biochem. J. **277**:33 – 38.

63. Jones T.W., Qin C., Schaeffer V.H., and Stevens J.L., 1988 Immuno–histochemical localization of glutamine transaminase K, a rat kidney cysteine conjugate 6–lyase, and the relationship to the segment specificity of cysteine conjugate nephrotoxicity. Mol. Pharmacol. **34**:621 – 627.

64. Elia M., and Livesey G., 1983 Effects of ingested steak and infused leucine on forelimb metabolism in man and fate of the carbon skeletons and amino groups of branched–chain amino acids. Clin. Sci., London. **64**:517 – 526.

65. Eaton S.B., Konner M., and Shostik M., 1988 Stone agers in the fast lane: Chronic degenerative diseases in evolutionary perspective. Am J. Med. **84**:739 – 749.

66. Cooper A.J.L., 1990 Ammonia metabolism in normal and portacaval shunted rats. Adv. Exp. Med. Biol. 272: 23 – 46.

67. Wibom R., and Hultman E., 1990 ATP production rate in mitochondria isolated from microsamples from human muscle. Am. J. Physiol. **259**:E204 – E209.

68. Graham T.E., and MacLean D.A., 1992 Ammonia and amino acid metabolism in human skeletal muscle during exercise. Can J. Physiol. Pharmacol. **70**:132 – 141.

69. Felig P., 1973 The glucose – alanine cycle. Metabolism. **22**:179 – 207.

70. Eriksson L.S., Broberg S., Bjorkman O., and Wahren J., 1985 Ammonia metabolism during exercise in man. Clin. Physiol. **5**:325 – 336.

71. Ganda O.P., and Ruderman N.B., 1976 Muscle nitrogen metabolism in chronic hepatic insufficiency. Metabolism **25**:427 – 435.

72. Dejong C.H.C., Kampman M.T., Deutz N.E.P., and Soerters, P.B., 1992 Altered glutamine metabolism in rat portal drained viscera and hindquarters during hyperammonemia. Gastroenterol. **102**:936 – 948.

73. Hundal H.S., Rennie M.J., and Watt P.W., 1987 Characteristics of L–glutamine transport in perfused rat skeletal muscle. J. Physiol. **393**:283 – 305 .

74. Bulus N., Cersosimo E., Ghishan F., and Abumrad N.N., 1989 Physiologic importance of glutamine. Metabolism: 38 Suppl. **1**: – 5.

75. Nelson D., Rumsey W.L., and Erecinska M., 1992 Glutamine metabolism by heart muscle. Properties of phosphate–activated glutaminase. Biochem. J. **282**:559 – 564.

76. Mudge G.H. Jr., Mills R.M. Jr., Taegtmeyer H., Gorlin R., and Lesch M., 1976 Alterations of myocardial amino acid metabolism in chronic ischemic heart disease. J. Clin. Invest. **58**:1185 – 1192.

77. Duffy, T.E., Plum, F., and Cooper, A.J.L., 1983 Cerebral ammonia metabolism in vivo. In: Glutamine, Glutamate and GABA in the Central Nervous System, edited by L. Hertz, E. Kvamme, E.G. McGeer, and A. Schousboe, Alan R. Liss, Inc., New York, pp. 371 – 381.

78. Hoyer S., Nitsch R., and Oesterreich K., 1990 Ammonia is endogenously generated in the brain in the presence of presumed and verified dementia of Alzheimer type. Neurosci. Lett. **117**:358 – 362.

79. Oldendorf W.H., 1971 Brain uptake of radiolabeled amino acids, amines, and hexoses after arterial injection. Am. J. Phsiol. **221**:1629 – 1639.

80. Smith Q.R., Momma S., Aoyagi M., and Rapaport S.I., 1987 Kinetics of neutral amino acid transport across the blood–brain barrier. J. Neurochem. **49**:1651 – 1658.

81. Abdul–Ghani A.–S., Marton M., and Dobkin J., 1978 Studies on the transport of glutamine in vivo between the brain and the blood in the resting state and during afferent electrical stimulation. J. Neurochem. **31:**541 – 546.

82. Welbourne T.C., 1988 Role of the lung in glutamine homeostasis. In: New Aspects of Renal Ammonia Metabolism. Contribution to Nephrology. Vol. 63, edited by G.M. Berlyne and S. Giovannetti, Karger, Basel, pp. 178 – 182.

83. Ardawi M.S.M., 1988 Skeletal muscle glutamine production in thermally injured rats. Clin. Sci. **74:**165 – 172.

84. Ardawi M.S.M., 1990 Glutamine metabolism in skeletal muscle of glucocortic–oid–treated rats. Clin. Sci. **79:**139 – 147.

85. Golden M.H.N., Jahoor P., and Jackson A.A., 1982 Glutamine production rate and its contribution to urinary ammonia in normal man. Clin. Sci. **62:**299 – 305.

86. Marliss E.B., Aoki T.T., Pozefsky T., and Cahill G.F. Jr., 1971 Muscle and splanchnic glutamine and glutamate metabolism in postabsorbtive and starved man. J. Clin. Invest. **50:**814 – 817.

87. Schröck H., and Goldstein L., 1981 Interorgan relationships for glutamine metabolism in normal and acidotic rats. Am. J. Physiol. **240:**E519 – E525.

88. Windmueller H.G., and Spaeth A.E., 1975 Intestinal metabolism of glutamine and glutamate from the lumen compared to glutamine from the blood. Arch. Biochem. Biophy. 171: 662 – 6672.

89. Windmueller H.G., and Spaeth A.E., 1980 Respiratory fuels and nitrogen metabolism in vivo in small intestine of fed rats. J. Biol. Chem. **255:**107 – 112.

90. Neptune E.M. Jr., and Mitchell T.G., 1964 Sodium and water transport by the everted rabbit intestinal ileum sac. Fed. Proc. **23:**152 abstract.

91. Fine A., 1982 The effects of chronic metabolic acidosis on liver and muscle glutamine metabolism in the dog. Biochem. J. **202:**271 – 273.

92. Yamamoto H., Aikawa T., Matsutaka H., Okuda T., and Ishikawa E., 1974. Interorgan relationships of amino acid metabolism in fed rats. Am. J. Physiol. 226: 1428 – 1433.

93. Kvamme E., Torgner I.Aa, and Roberg B., 1992 Evidence indicating that pig renal phosphate–activated glutaminase has a functionally predominant external localization in the inner mitochondrial membrane. J. Biol. Chem. **266:**13185 – 13192.

94. Windmueller H.G., and Spaeth A.E., 1976 Metabolism of absorbed aspartate, asparagine and arginine by rat small intestine in vivo. Arch. Biochem. Biophys. **175:**670 – 676.

95. Windmueller H.G., and Spaeth A.E., 1981 Source and fate of circulating citrulline. Am. J. Physiol. **241:**E473 – E480.

96. Taylor P.M., and Rennie M.J., 1987 Perivenous localisation of Na–dependent glutamate transport in perfused liver. FEBS Lett. **221:**370 – 374.

97. Häussinger D., and Sies H., 1979 Hepatic glutamine metabolism under the influence of the portal ammonia concentration in the perfused rat liver. Eur. J. Biochem. **101:**179 – 184.

98. Häussinger D., Gerok W., and Sies H., 1983 Regulation of flux through glutaminase and glutamine synthetase in isolated perfused rat liver. Biochim. Biophys. Acta **755:**272 – 278.

99. Häussinger D., 1989 Glutamine metabolism in the liver. Overview and current concepts. Metabolism 38: Suppl. **1:**14 – 17.

100. Buttrose M., McKeller D., and Welbourne T.C., 1987 Gut – liver interaction in glutamine homeostasis: Portal ammonia role in uptake and metabolism. Am. J. Physiol. **252:**E746 – E750.

101. Lenzen C., Soboll S., Sies H., and Häussinger D., 1987 pH control of hepatic glutamine degradation. Role of transport. Eur. J. Biochem. **166:**483 – 488.

102. Szweda L.I., and Atkinson D.E., 1990 Response of rat liver glutaminase to pH, ammonium and citrate. Possible regulatory role of glutaminase in ureagenesis. J. Biol. Chem. **265:**20869 – 20873.

103. Atkinson D.E., and Bourke E., 1984 The role of ureagenesis in pH homeostasis. TIBS **9:** 297 – 300.

104. Atkinson D.E., and Bourke E., 1987 Metabolic aspects of the regulation of systemic pH. Am. J. Physiol. **252:**F947 – F956.

105. Häussinger D., Kaiser S., Stehle T., and Gerok W., 1990 Stuctural and functional organization of hepatic ammonia metabolism: Pathophysiological consequences. In: Advances in Ammonia

Metabolism and Hepatic Encephalopathy, edited by P.B. Soeters, J.H.P. Wilson, A.J. Meijer, and E. Holm, Excerpta Medica, Amsterdam, pp.26 – 36.

106. Häussinger, D., 1990 Nitrogen metabolism in normal and cirrhotic liver. Adv. Exp. Med. Biol. **272**:47 – 64.

107. Walser, M., 1986 Roles of urea production, ammonium excretion, and amino acid _oxidation in acid – base balance. Am. J. Physiol. 250: F181 – F 188.

108. Cohen N.S., Kyan F.S., Kyan S.S., Cheung C.W., and Raijman L., 1985 The apparent Km of ammonia for carbamyl phosphate synthetase ammonia in situ. Biochem. J. **229**:205 – 211.

109. RÚmÚsy C., DemignÚ C., and Fafournoux P., 1986 Control of ammonia distribution ratio across the liver cell membrane and of ureogenesis by extracellular pH. Eur. J. Biochem. **158**:283 – 288.

110. Cheema–Dhadli S., Jungas R.L., and Halperin M.L., 1987 Regulation of urea synthesis by acid – base balance in vivo: Role of ammonia concentration. Am. J. Physiol. **252**:F221 – F221.

Clinical Manifestations and Therapy of Hepatic Encephalopathy

Juan Rodés

1. Introduction

Hepatic encephalopathy (HE) is a functional and generally reversible alteration of the central nervous system that appears in patients with acute and chronic liver disease. The pathogenesis of HE is not well understood (this aspect is deeply discussed in another chapter of this issue) although it has been suggested that the key mechanism is due to the inability of the liver to eliminate endogen and exogen products which are toxic to the brain. These substances reach the systemic blood stream because of the presence of hepatic failure, and the development of portosystemic shunts. Both mechanisms allow the intestinal blood to reach the brain without being cleared by the liver (1).

2. Clinical manifestations

Classically it has been considered that there are two forms of HE, acute and chronic (1). These two forms are, in turn, subdivided into four types: 1) acute or subacute HE, 2) acute or subacute recurrent HE, 3) chronic recurrent HE and, 4) permanent chronic HE (this type also includes patients with myelopathy).

The main causes of acute or subacute HE are severe acute hepatic failure and, less frequently, a few cases of hepatic cirrhosis or other chronic liver diseases. The second type includes those patients with more than one episode of HE, separated from each other by a relatively long period of time. In these patients cerebral function remains normal in the periods of time free of HE. The third type includes patients with recurrent HE, that are relatively easy to control under appropriate therapy. Finally, the fourth type, includes patients with chronic encephalopathy in which the cerebral function oscillates from time to time. In these patients there usually are histological cerebral lesions which, in some patients may be in the spinal medulla. This kind of HE usually develops in patients submitted to portosystemic shunts. This clinical picture is also termed non–wilsonian hepatocerebral degeneration (2).

In the last few years particular attention has been paid to the subclinical or latent form of HE (3, 4). This clinical condition may be quite frequent in patients with cirrhosis and its recognition is very important. In fact, these patients may present a lack of memory and alterations in the psycometric tests, consequently they may develop alterations in neuromuscular coordination, impairing driving ability (5).

The symptoms and signs of HE are very variable and two types may be considered: 1) alteration of the mental status and personality and, 2) neuromuscular alterations (1). In the

Liver Unit, Hospital Clínico y Provincial, University of Barcelona, Spain

Cirrhosis, Hyperammonemia, and Hepatic Encephalopathy,
Edited by S. Grisolia and V. Felipo, Plenum Press, New York, 1994

Table 1. Grading of hepatic coma (10)

Grade 0	No abnormality detected.
Grade 1	Trivial lack of awareness, euphoria or anxiety. Shortened attention span. Impairment in performing addition or substraction.
Grade 2	Lethargy or apathy. Disorientation for time. Obvious personality changes, inappropriate behaviour.
Grade 3	Somnolence to semistupor, but responsive to stimuli. Confused. Gross disorientation.
Grade 4	Coma. Mental state not testable.

acute and recurrent types of HE it is relatively easy to recognize four clinical stages (Table 1). This grading is based on changes in conscious intellectual function and behaviour but does not include neurological changes or asterixis. In stages III and IV, the Glasgow coma scale (Table 2) is an additional useful clinical parameter.

Table 2. Glasgow Coma Scale

Eyes	Open	Spontaneously	4
		To verbal command	3
		To pain	2
		No response	1
Best motor response	To verbal command	Obeys	6
	To painful stimulus	Localizes pain	5
		Flexion–abnormal (decorticate rigidity)	3
		Extension (decerebrate rigidity)	2
		No response	1
Best verbal response		Oriented and converses	5
		Disoriented and converses	4
		Inappropriate words	3
		Incomprehensible sounds	2
		No response	1
Total			3–15

The clinical picture of HE starts slowly although in some cases HE initiates suddenly. Initially, there are subtle behaviour and character disturbances and disorders of sleep pattern. In other ocassions the patients develop psychological alterations with mild confusion, euphoria or depression. Later drowsinees, lethargy, gross deficits in ability changes and intermitent disorientation may be detected. In deeper cases the patients are somnolent but rousable and are unable to perform mental tasks. Disorientation, marked confusion, amnesia, occasional fits of rage and speech, although incomprehensible, may also be present. Finally, the patients develop coma (6) (Table 2).

In non–comatose patients flapping tremor (asterixis) is always present. This sign is very characteristic of any type of HE although it may be observed in other clinical conditions (uremia, respiratory failure and hypokaliemia). This sign accordingly to McIntyre (7) is explained as follows: "ask the patient to raise both arms horizontally in front of him (palms downwards), to dorsiflex the wrists and spread the fingers wide apart, and to hold this posture for about 15 seconds. The term flap is used to describe the small, brief, intermittent movements of individual fingers, either in flexion or laterally in an ulnar direction, with a rapid return of the fingers to the original position. With more severe asterixis the flap spreads proximally, movements involve the wrist and even the shoulders, and in extreme cases you may see movements of the head if it is held erect". Asterixis is random and asymmetrical.

It is not due to contraction, but to sudden relaxation of muscles; electromyograms from the contracted muscles holding the posture show frequent short periods of reduced electrical activity, which may account for the tremulous. Asterixis occurs when longer periods of absolute silence coincide in different muscles. Asterixis has been included by grading HE (Table 3). There are other clinical manifestations related to neuromuscular disfunction which include a lack of an appropriate coordination of several muscular groups (this trouble may be explored inviting the patient to write or draw simple geometrical figures), changes in the muscular tone, hyperreflexia and focal or generalized convulsions. In a very few cases spastic myelopathy may be detected (7).

Finally there are two symptoms not related to alterations of the central nervous system hyperventilation and foetor hepaticus (6).

Table 3. Portosystemic encephalopathy index (10)

EEG (cycle/sec)		Number connection test (sec)	
Normal	0 points	30	0 points
8.5 – 12	1 points	31 – 50	1 points
7 – 8	2 points	51 – 80	2 points
3 – 5	3 points	81 – 120	3 points
3	4 points	120	4 points
Flapping tremor		Arterial ammonia (ug/dl)	
Absent	0 points	150	0 points
Isolated	1 points	150 – 200	1 points
Irregular	2 points	201 – 250	2 points
Frequent	3 points	251 – 300	3 points
Constant	4 points	300	4 points

3. Diagnosis

The diagnosis of HE is bassed on the above mentioned clinical data although it is important to take into account that cirrhotic patients may present these clinical manifestations due to causes other than HE such as: meningitis, cerebral hemorrhage or cerebral hematoma.

Electroencephalogram (EEG) may be useful in the diagnosis of HE although the alterations observed may also be seen in other diseases. The changes observed in the EEG consist in a decrease in wave frequency and an increase in wave amplitude. This means that instead of a normal alfa rythmn of 8–13 cycles/seg a the theta rhytmn of 5–7 cycles/seg appears. In more advanced cases delta waves may be detected (1). In patients with deep coma progressive flattening of the waves may be seen. By using a computerized system it is possible to calculate the mean dominant EEG frequency which allows the quantification of the EEG alterations (8).

The visual evoked potentials is a relatively new method introduced in the evaluation of HE. Initially, it was considered that this method may be more useful than EEG; however, at present, it has been demonstrated that its clinical value is similar to EEG (9).

The number connection test is the best psychometrical method to detect and follow–up HE. This test consists in the establishment of the time required by the patient to follow the number (from 1 to 25) randomly distributed on a sheet of paper (10). Blood ammonia measurement may be of interest in the diagnosis of HE, particularly if it is related to other clinical and EEG data. Conn described the PSE index (Portosystemic encephalopathy index) to quantify the HE grade (Table 3). This index is particularly useful to evaluate the effect of a given therapeutic schedule (1,10).

4. Therapy

Treatment of HE depends on several factors, the most important being the kind of HE (acute or chronic forms), the importance or deepness of neurological disorders and the presence and cause of precipitating factors.

The general measures that should be taken into account are identical to those of any patient with a diminished level of consciousness such as: general nursing care, control of electrolytes and fluids, arterial pressure, heart and respiratory rates, temperature and diuresis. It is very important to pay particular attention to the development of bacterial infections (urinary and respiratory) and respiratory and renal function.These patients should be treated in an intensive care unit, particularly those with grade III and IV HE or with acute hepatic failure (in this overview the treatment of acute hepatic failure is not considered).

In patients with acute hepatic encephalopathy possible precipitating factors should be sought and treated appropriately: correction of hydroelectrolyte disturbances, treatment of gastrointestinal bleeding, antibiotic therapy when there are bacterial infections, correction of constipation, and avoidanse of the administration of sedative substances and diuretics. Moreover, all the measures destined to the diminishing of the synthesis or intestinal absorption of nitrogen compounds are essential in the management of HE. In this sense diet should be free of proteins, and intestinal washout performed. Drugs able to diminish the intestinal bacterial flora responsible for ammonia synthesis such as oral non absorbable antibiotics (neomycin and paramomycine), which produce a reduction of proteolytic bacterias may be used (10). However, the most common drug used in HE is lactulose. This drug is a disaccharide that is not metabolized by the intestine since there is no disaccharidase in the intestinal mucosa of human beings. Consequently, when lactulose reaches the colon it is hydrolized by sacarolitic bacteria. The exact mechanism by which lactulose improves HE is, at present, not well understood. It is considered that its beneficial effect is related to the fact that, in addition to producing an osmotic laxative effect due to hydrolysis products, it also carries out an acidification in the intestinal lumen which, in turn, diminishes the intestinal absorption of the ammonium. These two mechanisms reduce the intestinal absorption of nitrogen compounds to the systemic circulation (11). Some patients do not tolerate lactulose because of the excessive sweet taste and meteorism. Furthermore, lactulose has been described to produce hypernatremia secondary to its excessive laxative effect in patients treated chronically with this drug (12). In the last few years a new drug, lactitol, similar to lactulose, has been introduced in the armamentarium of the management of HE. The therapeutic efficacy of this drug is similar to lactulose although it has several advantages since the taste is less sweet and the incidence of meteorism is lower than lactulose (13).

It has been considered that the intravenous administration of branched amino acids may be very useful in the management of patients with acute HE (14). However, several control clinical trial were unable to demonstrate that real efficacy of this treatment in patients with acute HE (4,15). Similarly, other treatments: L–dopa, bromocriptine and flumazenil (benzodiazepine antagonist) were unable to demonstrate their usefulness in the management of these patients (16).

In summary, the treatment of acute HE consists in: the identification and treatment of the precipitating factors, restriction of protein ingestion and administration of neomycin, lactulose or lactitol. The dose of neomycin is 1 gr/6 h., whilst the dose of lactulose and lactitol varies from patient to patient. Dosification should be adjusted according to the number of bowel movements (2-3 bowel movement per day is considered the best response to be obtained with lactulose or lactitol). The association of neomycin and lactulose may produce variable effects: synergism or antagonism, therefore this association should be avoided particularly as an initial treatment (10).

When the acute episode of HE is cured it is essential to increase protein content in the diet to maximum tolerance and to maintain the treatment with lactulose or lactitol to avoid constipation and obtain at least two bowel movements per day.

The treatment of chronic HE (recurrent or persistent) may present subtle differences (lactulose or lactitol are the drugs indicated to control these patients) (17). The chronic administration of neomycine should be avoided in these patients since it may produce ototoxicity and nephrotoxicity through a minimal but constant intestinal absorption of the antibiotic. Protein intake should be reduced according to individual tolerance; in some cases this may be very low and may produce nutritional problems. This aspect is solved by increasing the ingestion of vegetable proteins. The use of other drugs such as: L–dopa and bromocriptine have no shown any therapeutic advantage.

In patients with surgical portosystemic shunt with chronic HE, supression of the shunt (surgical ligation or balloon occlusion) may be useful to improve chronic HE, however in these circumstances the risk of gastrointestinal bleeding due to a rupture of esophageal varices is very high (18). Finally, in selected patients, liver transplantation may be an excellent therapy in patients with chronic HE (19). In Table 4 there is a summary of HE therapy.

Table 4. Treatment of PSE.

Acute episodes of PSE

1. Elimination of precipitating factors.
   – Empty bowel of blood.
   – Cessation of diuretics and restoration of fluid and electrolyte balance.
   – Treatment of infections.
   – Low protein intake (40 g/day).
   – Avoidance of sedative drugs.
2. Drug treatment.
   – Lactulose, lactitol or neomycin.
   – Flumazenil if history of benzodiazapine administration (otherwise its use is experimental).

Chronic PSE

1. Diet.
   – Protein intake of 60 g/day or more.
   – Lactovegetarian diet preferable.
   – Only in severe protein intolerance replace part of oral proteins by branched chain amino acids.

2. Drugs.
   – Nonabsorbable disaccharide in doses sufficient to produce two soft bowel movements/day.
   – Diuretics reduced to a minimum.
   – Sedatives to be avoided
   (Exception: small doses of oxazepam if absolutely needed).

Recently, the administration of carbamylglutamate in the treatment of hyperammoniemia states has been re–evaluated by using an experimental model. The administration of carbamylglutamate produced a marked decrease of ammonia levels in blood and the excess of ammonia was eliminated in urine as urea. These results suggests that oral administration of carbamylglutamate may be useful in the treatment of hyperamoniemia states, including enzymatic deficiences and in the most frequent pathological substances of cirrhotic patients (20). However, to know whether carbamylglutamate may be useful in the management of patients with HE, double–blind control clinical trials are required.

References

1. Fraser, C. L., and Arieff, A. I., 1985, Medical progress: hepatic encephalopathy, New Engl J Med. **313:**865– 873.
2. Victor, M., Adams, R. D., and Cole, M., 1965, The acquired (non–Wilsonian) type of chronic hepatocerebral degeneration, Medicine (Baltimore). **44:**345–396.
3. Rikkers, L., Jenko, P., Rudman, D., and Freides, O., 1978, Subclinical hepatic encephalopathy:

detection, prevalence and relationship to nitrogen metabolism, Gastroenterology. **75**:462–469.

4. Tarteer, R. E., Hegedns, A. M., and VanThiel, D. H., et al, 1984, Non–alcoholic cirrhosis associated with neuropsychological disfunction in the absence of overt evidence of hepatic encephalopathy, Gastroenterology **82**:1421–1427.

5. Schomens, H., Hamster, W., and Blunck, H., et al, 1981, Latent portosystemic encephalopathy. I. Nature of cerebral functional defects and their effect on fitness to drive, Dig Dis Sci **26**:622–30.

6. Ferenci, P., Hepatic encephalopathy. Ed. McIntyre, N., Benhamou, J.P., Bircher, J., Rizzetto, M., and Rodés, J., 1991, En: Oxford Textbook of Clinical Hepatology, Oxford. pp 471–484.

7. McIntyre, N., 1991, Symptoms and signs of liver diseases. In: Oxford Textbook of Clinical Hepatology. Ed. McIntyre, N., Benhamou, J. P., Bircher, J., Rizzetto, M., Rodés, J., Oxford. pp. 271–290.

8. Terés, J., Silva, G., Masana, J., and Rodés, J., 1987, Spectral analysis of the electroencephalogram predicts portal systemic encephalopathy in cirrhotic patients submitted to a distal spleno–renal shunt (abstract), J. Hepatol. 5 (suppl):66.

9. Zenerolli, M. L., Pinelli, G., and Gollini, G., 1984 et al. Visual evokade potentials: a diagnostic tool for the assessment of hepatic encephalopathy, Gut. **25**:291–299.

10. Conn, H. O., and Lieberthal, M. H., 1979, The hepatic coma syndromes and lactulose, Williams and Wilkins. Baltimore.

11. Sherlock, S., 1987, Chronic portal systemic encephalopathy: update 1987, Gut. **28**:1043– 1048.

12. Nelson, D. C., McGraw, W. R. G. Jr., and Horympa, A. M. Jr., 1983, Hypernatremia and lactulose therapy. JAMA. **249**:1295–1298.

13. Heredia, D., Caballería, J., Arroyo, V., Ravelli, G., and Rodés, J., 1987, Lactitol versus lactulose in the treatment of acute portal systemic encephalopathy (PSE), A controlled trial, J. Hepatol. **4**:293–2981.

14. Rossi–Fanelli, F., Riggio, O., and Caniano, C., 1982, et al. Branched chain aminoacids vs lactulose in the treatment of hepatic coma: a controlled study, Dig Dis Sci. **27**:475–480.

15. Morgan, M. Y., 1990, Branched chain aminoacids in the management of chronic liver disease. Facts and fantasies, J. Hepatol. **11**:135–141.

16. Bansky, G., Meier, P. J., Riedered, E., Walser, H., Ziegler, W. H., and Smid, M., 1989, Effects of the benzidiazepine receptor antagonist flumazenil in hepatic ecephalopathy in humans, Gastroenterology. **97**:744–750.

17. Morgan, M. Y., Hawley, K. E., and Strambuck, D., 1987, Lactitol vs lactulose in the treatment of chronic hepatic encephalopathy: a double–blind randomized, crossover study, J. Hepatol. **4**:236–245.

18. Potts, J. R., Henderson, J. M., and Millikan, W. J., Jr., 1984, et al. Restoration of portal venous perfusion and reversal of encephalopathy by balloon occlusion of portal–systemic shunt, Gastroenterology. **87**:208–212.

19. Bismuth, H., Samuel, D., and Gugenheim, J., et al, 1987, Emergency liver transplantation for fulminant hepatitis, Ann Int Med. **107**:337–341.

20. Grau, E., Felipo, V., Miñana, M. D., and Grisolia, S., 1992, Treatment of hyperammoniemia with carbamylglutamate in rats, Hepatology. **15**:446–448.

21. Bircher, J., and Sommer, W., 1992, Portal–systemic encephalopathy. In: Hepatobiliary diseases. Ed. Prieto, J., Rodés, J., and Shafritz, D. A., Springer–Verlag. Berlin Heidelberg. pp 417–426.

Nutritional Considerations in Patients with Hepatic Failure

A. Sastre Gallego, E. Morejón Bootello, P. Carda Abella

1. Introduction

The evolution of the patients affected by chronic hepatopathy is a problem which results in a high percentage of surgical interventions. Portal hypertension leads to one of the most serious dangers for the lives of these patients: upper digestive hemorrhage with its hemodynamic and metabolic sequels. The situation is dramatized by an ever – increasing circumstance: the ethylic origen of the hepatic lesion strikes in increasingly younger ages, or in the best of cases, in the middle stages of life. It must be pointed out that the consumption of pure ethanol "per capita" annually in Spain is 13 liters. The consumption of beer has risen to 50 liters; wine consumption remains at 60 liters and spirits of 40º at 7.5 liters per capita and year [1].

Even other aetiologies which select the liver as the target organ also develop their irreversible sequels in the ages of youth and maturity. The portalsystemic anastomoses which establish an end– to–side or side–to–side shunt between the porta and lower cava to relieve portal hypertension and to bypass the danger of hemorrhages of the esophagus have always been faced with another danger, somewhat lesser, and in some ways closely related to that represented by the hemorrhage: the post–operative metabolic problems. And so, the possibility of developing encephalopathy and fatal coma have reached figures superior to 40%. It is a danger which remains with the patient throughout his entire life.

The cirrhotic patient, undergoing or not a portalsystemic anastomosis, is faced with a double–edged danger: a nutritional intake by default inadequate to maintain reserves, to achieve positive balance and to attempt histic regeneration; or a protein intake excessive for his metabolic capacities which will lead to encephalopathy.

A high percentage of cirrhotic patients maintain a precarious nutritional state which oscilates between malnutrition of medium or severe intensity, measured by very diverse indexes and methods. The most frequent alteration is the loss of muscular mass in 92.7% of the patients and the correlation between nutritional state and the evolution of the hepatopathy [2].

The daily food intake is very poor and can be attributed to several diverse causes:

1) Anorexia
2) Alcohol consumption
3) Unappetizing diets
4) Predominance of carbohydrates with an impoverishment of proteins and vitamins.

Unidad de Nutrición Clínica y Dietética, Dpto. de Cirugia General y Digestiva, Hospital "Ramón y Cajal", Madrid, Spain

Cirrhosis, Hyperammonemia, and Hepatic Encephalopathy,
Edited by S. Grisolia and V. Felipo, Plenum Press, New York, 1994

In the study by Leevy [3], 49% of the cirrhotic patients have a vitamin deficiency which can be detected by plasmatic determinants. The most important deficiencies are folate (47%), thiamin, riboflavin, nicotinic acid and pyridoxine (25%). Vitamin B12, pantothenic acid, and biotin are low in 20% of the cases.

Liposoluble vitamins also suffer reductions, especially vitamins A–E–K. In patients with primary biliary cirrhosis, a deficiency of vitamin C has also been detected [4].

The nutritional problems of those patients affected by evolutive chronic hepatopathy can be grouped into two branches: digestive and metabolic.

2. Digestive Level

2.1 Anorexia

The origen is multifactoral. In alcoholic patients the rejection of the consumption of other nutrients is conditioned by the ingestion of highly caloric ethanol. To this we must add habitual gastritis with frequent nauseas. The diets programmed are poor in sodium and proteins and result insipid and unappetizing. And, lastly, the existance of high levels of free tryptophan and cerebral serotonin sedate the central appetite center.

2.2 Malabsorption

Steatorrhea is the most common symptom. Fifty percent of cirrhotic patients suffer steatorrhea of medium intensity (>10 grams of fat/24 hours), whether their hepatopathy originates with alcohol or not.

The reduction of biliary salt has been frequently demonstrated in the intestional light; at the same time, deficiencies appear in the absorption of long–chain fatty acids and in contrast, the absorption of medium–chain fatty acids (TCM) and D–Xylose are normal.

There are also iatrogenic influences. Neomycin is capable of altering the intestional ecology with a certain toxicity on the enterocytes, as well as precipating biliary salts and fatty acids contributing to steatorrhea [5].

Cholestyramine, used to aliviate pruritus in cholestasis, produces steatorrhea and diminishes the absorption of vitamins A, D, E, K in animals and man [6].

3. Metabolic Level

The liver is a great metabolic regulator of glycemia, establishing a balance between the synthesis of glycogen and glycolysis (absorption phase) and between glycogenolysis and gluconeogenesis (post–absorption phase).

More than 50% of the glucose absorbed at intestional level is used in the synthesis of glycogen and triglycerides. The deposits of hepatic glycogen are capable of maintaining the necessary rate for 24 hours. In states of fasting, the liver pumps glycemia by means of the mechanisms of glycogenolysis. When these reserve sources have been exhausted, it is used in gluconeogenesis at the expenses of the amonoacids used as substrates. Control of glycemia is mediated by substrate concentration, the hepatic nervous system, the zonal and functional heterogenity of the hepatocyte and the insulin/glucagon hormonal relationship. Insulin is fundamentally glucogenogenic; glucagon is glucogenolitic and gluconeogenic. In hepatopathies, hypo– and hiperglucemias can appear as an expression of multiple regulation disorders.

1. Hypoglycemia is usually associated with alcoholism and fasting. It can be induced by low hepatic reserves of glycogen and certain inhibitions of the gluconeogenesis

mechanisms: alcohol metabolism is carried out by means of oxidizing processes which lead to the formation of acetaldehyde (cytoplasm) and acetate (mitochondria). Alcoholdehydrogenase and aldehydehydrogenase enzymatic systems lead to a reduction of nicotinamide adenine dinucleotide (NAD) and to an increase of NADH. The influence of the latter decreases the formation of phosphoenolpyruvate at the expense of pyruvate via oxalacetate which is the fundamental step of gluconeogenesis [7].

The lactate/pyruvate quotient also increases and this hiperlactacidemia makes the excretion of uric acid in the distal tubule difficult, causing at times, hipervericemia and gouty crisis.

2. Hiperglycemia is more frequent and can be found in up to 80% of the chronic hepatopathies. And this number can be even more elevated in those patients operated on for portalsystemic anastomosis. It seems evident that the mechanisms of hypersecretion, insuficient break down, and Insulin resistence are associated in the etiopathogenia.

At the time of reestablishing nutritional support to the hepatic patient it must be remembered that the secretion of insulin as a answer to glucose is greater by oral means than by parental means even in patients with portalcaval shunt. This suggests an active function of the gastrointestinal hormones, since the stimulating role of the secretion of secretin, pancreozymin gastrin and g.i.p. is well–known [8].

Certain aminoacids like arginine stimulate the secretion of insulin, strengthened by cholecystokinin, pancreozymin and gastrin. Somatostatin, adrenalin and prostaglandin have a depressing effect.

4. Lipid Metabolism

The liver intervenes decisively in this system since it participates in the synthesis and secretion of lipoproteins, attraction and breakdown of the same, synthesis of cholesterol and the excretion of certain lipids. It also synthesizes lecithin cholesterol – acyltransferase (L.C.A.T.) which mediates the esterification of cholesterol.

A decline in the activity of triglyceridelipase and LCAT can be found in viral and alcoholic hepatopathies. In consequence, the levels of non–esterified triglycerides and cholesterol increase. Alcohol contributes to the increase of plasmatic triglycerides by elevating the hepatic synthesis of VLDL and by inhibiting lipoprotein–lipase.

The increase of the synthesis and the reduction in the breakdown of fatty acids in the liver is also a habitual disorder in chronic hepatopathy. Synthesis is stimulated by the increase of reduced nicotinamide–diphosphonucleotide (NADPH) while parallelly oxidation diminishes on increasing the NADPH/NADP$^+$ ratio.

The Apo B, E receptors discovered in 1973 by Goldstein and Brown, recognize the Apo B–100 (LDL) and the Apo–E (IDL-LDL). The activity of these receptors increases in fasting with the ion exchange resins, thyroxine, estrogens and the h.m.g. CoA– reductase inhibitors.

Diets rich in saturated fats and cholesterol diminish the number of Apo B, E receptors and the clearing of IDL–LDL.

On the other hand the possible hepatic receptors for HDL (cholesterol of the tissue to liver) increase their activity with diets rich in cholesterol and fats [11] [12].

In certain phases of hepato–cellular insufficiency, there can be abnormally composed LDL lipoproteins which are interpreted as IDL which have evolved badly because of a partial blockage of the receptors of the hepatic Apo B, E. In the very advanced stages of chronic hepatopathy, the synthesis of VLDL and LDL (TG and cholesterol) diminishes.

Once the lipoproteins are internalized in the hepatocyte, the lipids undergo breakdown and elimination through the biliary canaliculus. Two hundred to six hundred mg. of biliary salts are produced daily.

5. Protein Metalbolism

Among the most representative proteins of hepatic genesis are:

5.1 Albumin

It constitutes 25% of the total of the hepatic proteins and 50% of the total protein exported through the liver. It is in charge of maintaining 80% of the colloidosmotic pressure.

It has detoxifying functions and is a great reserve of aminoacids. Its half life is 20 days.

The liver produces 12 g daily and only a 10–20% of the hepatocytes are capable of maintaining the sufficient ammount of albuminemia.

Portal hypertension in the cirrhotic patient and a drop to below 3 g/100 ml in the figures of normality are conditioners of ascites.

5.2 Pre–albumin

It forms complexes with the protein together with retinol, 0.3g/1000 ml.

5.3 Alfa 1 antitrypsin

2.0 g/1000 ml. Can be diminished in cirrhosis.

5.4 Transferrin

2.5 g/1000 ml. It is usually diminished in patients with hepatopathy.

The coagulation factors, the complement C3 factor and the α–fetoprotein can be modified in cirrhosis and hepatocarcinoma.

The extrahepatic genesis proteins can also undergo variations in hepatopathies, fundamentally the immunoglobulins.

6. Ammonia, Urea and Glutamine Synthesis

One of the most important metabolic functions of the liver is the elimination of ammonium by two fundamental mechanisms:

a) urea synthesis
b) glutamine synthesis

Ninety–five percent of the levels of arterial ammonium maintained by the production in the large intestine, small intestine and in the kidney is eliminated by means of urea synthesis in the Krebbs–Henseleit cycle [13]. When this is altered as habitually happens in chronic hepatopathy, the synthesis of glutamine increases. Urea is irreversible but the glutamine, by action of the glutaminases can be reconverted into glutamate + ammonium. So there is no excess of glutamine. There seems to be a certain zonal division, interrelated, in the liver to adequately distribute these functions.

7. Vitamins and Minerals

In chronic hepatopathies several types of disorders can be detected in relation with micronutrients:

1. insufficient ingestion and malasorption
2. diminishing of hepatic storage
3. difficulty of conversion to active forms
4. increase in necessities

In hepatopathic patients, the hepatic reserves of folate, riboflavin, nicotinamide, pantothenic acid, vitamin B6, vitamin B12, and vitamin A diminish. In episodes of necrosis there is a clear loss of hydrosoluble vitamins. There is, besides, a lesser hepatic offer of vitamins in active form.

8. Vitamins A–D–E–K

These are steroid derivatives soluble in fat. The liver transforms the discontinuous intake of retinol into a continuous storage to pump into the rest of the tissues. The hydrolysis of retinol palmitate in the liver pushes the vitamin into the circulatory stream forming a complex with pre–albumin and P.U.R. In the tissues, retinol metabolizes to retinal and retinoic acid.

Low levels of vitamin A and symptoms of night blindness are frequently found in the cirrhotic patient. But certain caution is necessary when treating alcoholic hepatopathy with high doses of retinol. Due to the interaction of the alcohol in the metabolism of vitamin A, microsomal toxic lesions are produced in the hepatocyte.

It is necessary to pay attention to the levels of zinc because this oligioelement is essential for the synthesis of the proteins which transport retinol.

Vitamin D is metabolized to active form 25–hydroxyvitamin D in the liver. This metabolite becomes 1,25–hydroxyvitamin D at renal level which is its definitive active form.

In cirrhotic patients a form of osteoporosis is already known and is associated with low levels of 25–hydroxyvitamin D. When oral or parental treatment with vitamin D is not efficient 25–hydroxyvitamin D can be used. Nevertheless as remineralization is not an adequate response, more complex mechanisms can be deduced in this function [14].

Vitamin E is the most effective antioxidizer in nature. α–tocopherol combines with the free radicals in the membrane and prevents lipooxidation. The product of the oxidation of α–tocopherol fuses in the hepatocyte with glucuronic acid and is eliminated with the bile.

Vitamin K is an essential cofactor for carboxyglutamic – carboxylase which converts the glutamic residues into an aminoacid: carboxyglutamic acid. It acts as a chelate on the calcium ion and intervenes in the mechanisms of coagulation.

9. Hydrosoluble Vitamins

9.1 Thiamine

It seems that the cirrhotic liver not only has difficulty in the phosphorylation of thiamine and its change into the active form pyrophosphate of thiamine but is also incapable of its complete utilization in this form. Various works have demonstrated that the addition of thiamine to red blood cell hemolysis increases the action of the erythrocyte transcelolase in healthy subjects, but not in cirrhotic patients [15].

9.2 Folate

The metabolic conversion of absorbed folic acid (pteroilglutamic) to 5–methyltetrahydrafolic acid takes place in the liver. The active form is necessary for the synthesis of desoxiribonucleic acid. It is possible that alcohol favors its storage in the pteroilglutamic form, and prevents its biliary secretion and the entero–hepatic cycle for the formation of 5–methyltetrahydrafolic. Enteric recuperation also affects vitamin B12.

9.3 Vitamin B6

Vitamin B6 is ingested as pyridoxine. It should undergo a phosphorilization by kinase, then by oxidase becoming pyridoxal to be later emptied into the plasma in active form as phosphate of pyridoxal. The liver is the principal agent of the functions. Low plasmatic levels have been described in alcoholic cirrhosis as well as in other aetiologies.

All this seems to suggest that for the conversion into the active form an increase in the mechanisms of destruction can take part. Vitamin B1 as well as B6 and biotin actively participate in the three metabolisms. Thiamine intervenes in the metabolism of carbohydrates, via direct oxidation, and its lack accumulates lactate from pyruvate. Alcohol can inhibit the active absorption of thiamine. The sudden administration of carbohydrates to the alcoholic, without sufficient vitamin B1, can induce the symptoms of the Wernicke–Korsakoff syndrome. The conversion of tryptophan to niacin requires pyridoxal–phosphate (B6). These requirements can be increased in the metabolism and uptake of aminoacids (Transamination).

9.4 Riboflavin

Deficiency of this vitamin affects the metabolism of vitamin B6 since the oxidase that converts phosphate of pyridoxine into phosphate of pyridoxal is a flavoprotein.

10. Nutritional Consequences

There are two necessary parameters to take into account in the patient affected by chronic hepatopathy:

a) the intense catabolic state with protein–caloric malnutrition; b) protein intolerance with risk of encephalopathy.

The cirrhotic patient presents metabolic disorders that affect the three intermediate principles, but imbalance and development of the encephalopathic state is fundamentally marked by protein intolerance.

In this situation the nutritional therapeutic recourse, up to the decade of the 70s, was to maintain the patient with so–called "hypoprotein diets" which included scarcely 10–15 g of vegetable origen protein in 24 hours. The obliged norm of this situation is intense muscular catabolism, negative nitrogenous balance and an influx of aminoacids to the plasma to supply gluconeogenesis. Paradoxically, the aprotein diet can induce an invasion of the plasma by aromatic aminoacids capable of producing coma.

For a long time the aminoacid has been the only protagonist mentioned as the producing agent of hepatic encephalopathy. Its importance is evident but today it must be considered as only one more factor in the genesis of this complicated process of cerebral toxicity.

Around 1971 the school of Fischer proposed a new theory concerning the conditioners of encephalopathy. Under conditions of hepatic hypofunction or deviation of the blood to perihepatic level, the amines as well as their aminoacid precursors escape the metabolic inactivation of the liver and flood the aminergic periferic and central nervous systems. This avalanche of neutral aminoacids at the blood brain barrier conditions the decline in the cerebral synthesis of noradrenaline and dopamine with an increase of betahydroxiphenylethonelamine, octopamine and metabolites of indolamine including serotonin [16][17].

The flooding of the central nervous system by the amines which lead to encephalopathy is secondary to the disorders in the metabolism of the aminoacid precursors: phenylalanine, tyrosine and tryptophan. These neutral aminoacids compete with those branched chain aminoacids for one system of transport which controls the passage through the vessels of the blood brain barrier.

In 1979, James and col. suggested that glutamine, synthesized locally in the brain from ammonia, contributes to the massive incorporation of aromatic aminoacids from the blood, perhaps by means of an acceleration of transport. This theory tries to reach a metabolic unification between the classic theory in respect to ammonia toxicity and the new works relative to the invasion of the central nervous system by amines and neutral aminoacids as agents of encephalopathy.

In portalsystemic shunted rats, it can be proven that an increase in the transport of neutral aminoacids across the blood brain barrier appears starting from the first six hours after the operation and lasts the 60 days of the experience. Serotonin also increases in the first 6 hours and passes 16 mol/gm of brain at 11 days.

In contrast, the plasmatic decrease of branched chain aminoacids and the increase of the neutral ones only begins starting from the 2nd and 5th post–shunt day. However, in the brain there is an elevation of aromatic aminoacids from the first 6 hours and this situation lasts throughout the 60 days of the experience [30].

All this seems to suggest that there are complex mechanisms which govern the entrance of aminoacids in the central nervous system, not only the distortion of Fischer's Index or the Valine + Leucine + Isoleucine/Tyrosine + Phenylalanine relationship.

As far as the reversal of encephalopathy with mixtures of aminoacids (BCAA) and very low intake of aromatic aminoacids, Fischer's theory has been amply defended by some schools (Cerra [20], Rossi–Fanelli [21]) and contested by others (Warren [22]).

Perhaps the controversy could yield to a moderate position like that expressed by some authors such as Silk [23],[29].

1. Malnutrition in patients with chronic hepathopathy is evident.
2. This calorie–protein malnutrition is tied to very low immunity levels, frequent sepsis and increase in mortality.
3. Malnutrition and malasorption condition situations of electrolytic anomalies, infections and a tendency towards encephalopathy.
4. A good protein–energy intake is prescribed for the hepatic patient. If there is adequate tolerance, there is no reason for not using a normal protein intake or mixtures and conventional solutions by enteral or parenteral means.
5. If there is clear protein intolerance, with a danger of encephalopathy, branched chain aminoacid solutions should not be used exclusively, but rather mixtures with a decreasing supply of phenylalanine, tryosine, trytophan and methionine. At the same time, increasing doses of arginine and branched chain aminoacids are assigned.

Our experience with patients treated over a medium or long term after portalsystemic anastomosis is that the only way to achieve a sufficient nitrogenous intake and positive balance is to recur to mixtures enriched with branched chain aminoacids and very poor in neutral aminoacids [24].

The possibility of administering a high protein rate (70–90 g/day) together with an energy intake at the expense of other immediate principles (2000–3000 Kcal/day) is centered on the contribution of branched chain aminoacids (BCAA). This intake can be transported in mixtures by parenteral or enteral means, or by forming part of studied combinations of foods in an oral diet.

This energy intake by means of carbohydrates and lipids does not impose more problems than the tolerance expressed by the glycemia levels and the possible existance of hypertriglyceridemia.

The use of medium chain triglycerides (TMC) has proven useful since it eludes the problems of malasorption and energy intake by mitochondrial –oxidation without mediation of L–carnitine. On the other hand, the lesser hepatic availability of long chain fatty acids reduces the synthesis of lipoproteins (VLDL).

11. Therapeutic Steps

When a patient arrives at the hospital in a situation of decompensated hepatopathy, the following points have to be considered:

1. Emergency action for hepatic coma or pre–coma.
2. Short term treatment, substituting, as soon as possible, enteral means for the parenteral which offer greater risks in a deficient immunitary situation.
3. Long term maintainance program trying to avoid the danger of metabolic slumps which again lead to encephalopathy.

When faced with the first situation, the use of branched chain aminoacids in a 35% of the total nitrogenous intake with a reduction of phenylalanine and methionine seems recommendable.

Branched chain aminoacids have the following effects:

1. They stimulate protein synthesis at muscular level.
2. They can satisfy the energy requirements in catabolic state in the presence of the scarcity of ketone bodies because of reduced synthesis in hepatopathy.
3. At the blood brain barrier they seem to compete with aromatic aminoacids in regard to cerebral captation.
4. On diminishing catabolism, the endogenous production of ammonia is reduced with less presence of cerebral glutamine.
5. Recents works [25] establish that arginine (F.080=7.5%) [26] is an aminoacid which specifically stimulates protein synthesis through the synthesis of polyamines and is capable of elevating immunity levels.

12. Clinical Experience

In 1985 we published our first series in collaboration with the Department of Digestive Surgery on our hospital [27] on 22 cirrhotic patients operated for portal hypertension with portalsystemic anastomosis.

Ten patients presented hepatopathy of ethylic origen and 12 post–necrotic. The average was 53 years old (range from 28 – 67) and the hepatic function was classified

Table 1.

TOTAL PATIENTS 22			
AGE		**SEX**	
Range: 28 – 67		Male 12	
Mean: 53		Female 3	
ALCOHOLIC CIRRHOSIS 10			
POSTNECROTIC		12	
URGENCY	8	16	MESO – CAVA H
PROGRAMMED	14	6	WARREN
MORTALITY: 13.2 %			

according to the criterion of Child (8 persons were grouped in stage A, 12 in B, and 2 in C). Eight of the patients had emergency operation for massive hemorrhage and the rest in a programmed manner during the three weeks following the hemorrhage incident (Table 1).

All were treated individually using parenteral nutrition during the first 72 post operative hours according to necessities on an average of 25–35 Kcal/Kilo of weight per day.

The formula began with 1000 Kcal by means of a mixture of aminoacids according to Fischer and 20% of dextrose for a total average supply of 1800–2000 Kcal/day. After the fourth post–operative day, the patients received enteral nutrition through a nasal–gastric catheter, the formula being adapted to energetic necessities and following the postulates of Fischer. This diet was maintained for 15 days with an oscillating supply of 2500– 3000 Kcal/day.

Afterwards the patients were discharged with an oral diet calibrated following the postulates of Fischer with a selection of natural foods (60 g of protein – 45% animal origen and 55% vegetable; 350 g. carbohydrates and 100 g of vegetable fat). A group pf patients, 8 in all, received, besides, a dietary supplement of 25–30 g. of a mixture of branched chain aminoacids.

We also established a comparative study of 9 patients diagnosed with cirrhosis and hospitalized in the Gastroenterology Service for decompensation and treated with a conventional methodolgy (hypoprotein diet, calorie intake 1000–1200 Kcal/day, intestinal action antibiotics and lactulose).

We had previously established the plasmatic aminogram figures with a group of 21 persons, considered healthy, of both sexes and with an age range of between 25–50 years old (Fischer's Index 2, 5–3) (Table 2).

The mortality rate of the surgically operated group was 13.2% (3/22)due to hepatic failure (2 patients in the C classification of Child and 1 patient in the B).

All were emergency operations. One patient died of endocarditis six weeks after the operation. The remaining 18 patients recovered an excellent quality of life and were followed by means of revisions every 3 months for 3 years after the operation.

The only incidents, grade I–II encephalopathy, were presented by three patients who went off the prescibed diet and ingested alcohol. They recovered and no other complication was presented.

The follow–up parameters for these patients were:

1. Clinical evolution
2. Biochemical tests of hepatic function

Table 2.

VALINE 2.8 ± 0.22		ISOLEUCINE 0.78 ± 0.17		LEUCINE 1.54 ± 0.26
PHENYLALANINE 1 ± 0.22			TYROSINE 1 ± 0.24	
RATIO = 2.5 ± 0.25				

3. Plasmatic aminogram and Fischer's Index
4. Shunt permeability

In the first and second figures the evolution of the plasmatic aminoacids are appraised from the moment of the portalsystemic shunt and the post–operatory on long term with BCAA supplement and without it.

A positive nitrogenous balance is reached from the 2nd week of post–surgical treatment. The increase on the Fischer Index was seen from the first week of treatment (Fig. 1).

This increase is not significant before the third week of treatment. Nevertheless, there is a significant decrease of aromatic aminoacids in the plasma of the sugically intervened patients after the second week of treatment compared with the figures of the healthy control group ($p < 0.005$).

On the Index of Fischer there are significant differences between the group of cirrhotic patients treated with a surgical shunt and those not operated on and treated with conventional therapy as compared with the healthy control group (Fig. 1).

There is also a difference between the group of unoperated cirrhotic patients and that treated by means of a shunt and therapy with BCAA for 3 weeks, being more favorable the results of the latter (Fig.1).

The favorable evolution of the clinical symptoms has been shown to be parallel to the elevation on the Index of Fischer.

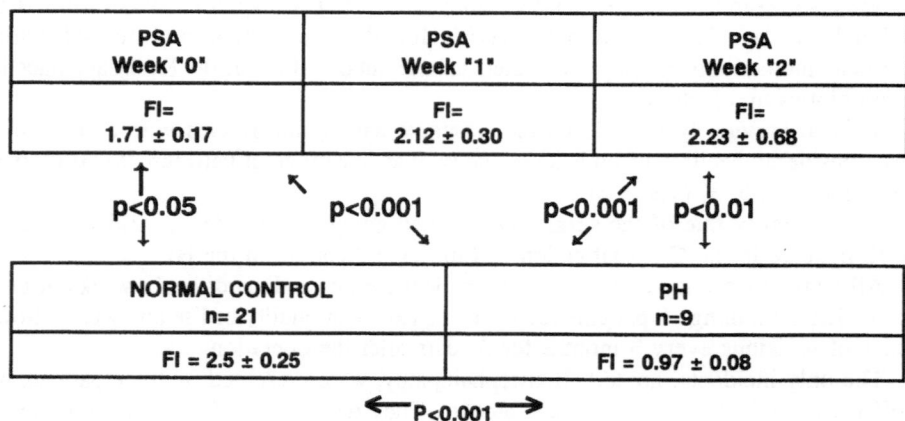

PSA = Portho–Systemic anastomosis.
PH = Porthal hypertension.
FI = Fischer Index.

Figure 1

13. Conclusions from our Work

1. The surgical derivations which maintain a certain hepatic flow are advisable even though there is no unification of criterion in this surgical field.

2. Nutritional support by means of enteral, parenteral and oral diets with BCAA has been shown to be efficient to check catabolism, maintain positive nitrogenous balance and provide a sufficient energy intake without danger of encephalopathy.

3. Those patients affected by cirrhosis to a relatively equal degree, and treated with conventional methodology, show a poorer nutritional state, a lower Fischer Index and a greater tendency toward encephalopathy.

4. On a long term, maintaining a Fischer Index above 2 by means of diets restricting aromatic aminoacids and an intake of BCAA, including 25–30 g/day of BCAA as a supplement has been shown to be efficient to maintain an adequate nutritional state and prevent encephalopathy (Fig.2).

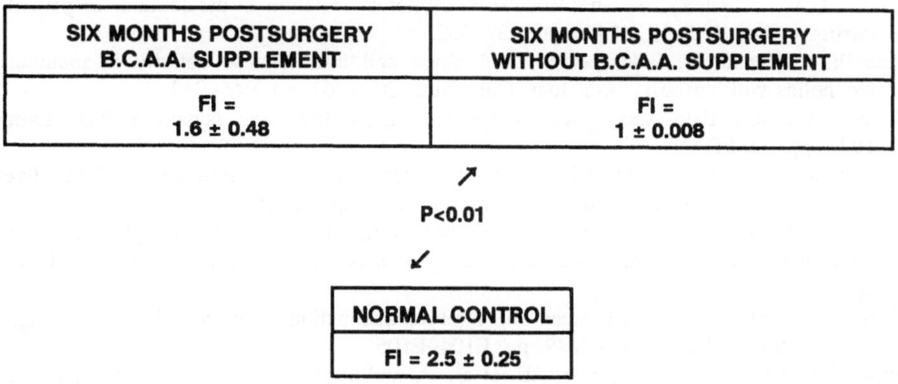

SIX MONTHS POSTSURGERY B.C.A.A. SUPPLEMENT	SIX MONTHS POSTSURGERY WITHOUT B.C.A.A. SUPPLEMENT
FI = 1.6 ± 0.48	FI = 1 ± 0.008

↗

P<0.01

↙

NORMAL CONTROL
FI = 2.5 ± 0.25

Figure 2

We have continued a follow–up of a group of 9 patients for 10 years. All have had an acceptable quality of life without important intercurrent episodes. Nevertheless, from 8–9 years after surgery, the hepatic situation has iniciated its definitive decompensation. We can affirm that in our series 86.5% have died in this period even with an excellent control and treatment of the disease. Of them, 25% have developed a hepatoma on the pre–existing cirrhosis.

References

1. Caballeria, J. and Pares, A., Alcohol e hígado en enfermedades digestivas. Ediciones CEA, Madrid, 1990, pp. 2103–2117.
2. Pomar, M.J., Bruguera, M., Caballeria, J., and Rodes, J., Malnutrición calórico–proteíca en pacientes hepáticos hospitalizados. Gastroenterología y Hepatología, Vol. 10, nº 9, 1987, pp. 434–440.
3. Leevy, C.M., Baker, H., Hove, W., et al., P–complex vitamins in liver disease of the alcoholic. Am. Jour. of Clin. Nutr. 16, 1965, pp. 339–346.
4. Mezey, E., Liver disease and nutrition. Gastroenterology, 74, 1978, pp. 770–783.

5. Thompson, G.R., Barrowman, J., Gutierrez, L., et al. Actions of Neomycin on the intraluminal phase of lipid absorption. J. Clin. Invest., 50, 1971, pp. 319–323.

6. Hashim, S.A., Bergen, S.S., Van Itallie, T.B., Experimental esteatorreha induced in man by bill [sic] acid sequestrant. Proc. Soc. Exp. Biol. Med. 106: 1961, pp. 173–175.

7. Collins, j.R., Lacy, W.W., and Stiel, J.N. et al. Glucose intolerance and insulin resistence in patients with liver disease. II A Study of aetiologic factors and evaluation of insulin actions. Arch. Inter. Med. 127, 1970, pp. 608–614.

8. Dupre, J., Curtis, J.D., Unger, R.H. et al. Effects of secretin, pancreozymin or gastrin on the response of the endocrine pancreas to administration of glucose or arginine in man.

9. Goldstein, J.L., Kita, T., and Brown, M.S., Defective lipoprotein receptors and atherosclerosis, lessons from an animal counterpart of Familial Hipercolesterolemia.

10. Brown, M.S., Kovanen, P.T. and Goldstein, J.L., Regulation of plasma cholesterol by lipoprotein. Science 212, 1981, pp. 628–635

11. Hoeg, J.M., Demosky, S.J., and Greff, R.E., Distinct hepatic receptors for low density lipoprotein and apolipoprotein E in humans. Science 277, 1985, pp. 759–761.

12. Manley, R.W. and Interarity, J.L., Lipoprotein receptors and cholesterol homeostasis. Biochem. Biophys. Acta 737, 1983, pp. 197–222.

13. Haussinger, D., Hepatocyte heterogeneity in glutamine and ammonia metabolism and the role of an intercellular glutamine cycle during ureogenesis in perfused rat liver. Am. J. Biochem. 133, 1983, pp. 269–275.

14. Hepner, G.W., Roginsky, M. and Moo, H.F., Abnormal vitamin D metabolism in patient with cirrhosis. Am. J. dis. 21, 1976, pp. 527–532.

15. Fennelli J., Frank, O., Baker, et al., Red blood cell transketolase activity in malnourished alcoholics with cirrhosis. Am. Jour. Clin. Nutr. 20, 1967, pp. 946–949.

16. Fischer, J.E. and Baldessarini, R.J., False neurotransmitters and hepatic failure. Lancet 2, 1971, pp. 75–85.

17. Baldessarini, R.J. and Fischer, J.E., Serotonin metabolism in rat brain after judicial diversion of the portal venoces circulation. Nature 245, 1973, pp. 25–35.

18. Hames, H.J., Jepson, B., Ziparo, V., Fischer, J.E., Hyperammonemia plasma aminoacid imbalance, and blood brain aminoacid transport: a unified theory of portal. Lancet ii, 1979, pp. 772.

19. Mans, A., et al., Early establishment of cerebral disfunction after portacaval shunting. Am. Physiol. Soc., 1990, 0193 – 1849, pp. E104–E106.

20. Cerra, F.B., Cheung, N.K., Fischer, J.E. et al., A multicenter trial of branched chain enriched aminoacid infusiores (F080) in hepatic encephalopathy. Hepatalolgy Z., 1982, pp. 699.

21. Rossi-Fanelli, F., Cascino, A., Cangiano, C., Branched chain aminoacids in the management of hepatic encephalopathy. J. Clin. Gastroenterol. Z., 1987, pp. 44–46.

22. Warren, K.S., Shenker, S., Effect of and inhibitor of Glutamine Synthesis on Ammonia toxicity and metabolism. J. Lab. Clin. Med. 64, 1964, pp. 442.

23. Silk, D.B.A., Parenteral nutrition in patients with liver disease. Journal of Hepatology 7, 1988, pp. 269–277.

24. Sastre, A., et al., Nutrición en la insuficiencia hepática. In Nutrición artificial en el paciente grave. Edit. Doyma. Barcelona, 1989, pp. 45–47.

25. Reynolds, J.V., Daly, J.M., et al., Immunomodulatory mechanisms of arginine. Surgery, 104, nº 2, 1988, pp. 143–151.

26. Barbul, A., Lazarou, S.A., Efron, D., et al., Arginine entrances wound healing and lynphocyte immune responses in humans. Surgery, Vol. 108, nº 2, 1990, pp. 331–337.

27. Carda, P., Saastre, A., La Roche, F., Morales, V. and Marcos, J. M. Correction of Aminoacid imbalance in Postoperative portalsystemic Shunts. Infusiontherapie 12, 1985, pp. 251–253.

28. Garcia Almansa, A., et al., Nutrición y hepatopatía, In Introducción a la Nutrición Clínica hospitalaria. P.P. Garcia Luna, Edit. SAS, 1990, pp. 251–260.

29. Muñoz, S.J., Nutritional therapies in liver disease. Seminars in liver disease, Vol. II, nº 4, 1991, pp. 278–291.

30. Mans, Anke M. et al., Early establishment of cerebral dysfunction after portacaval shunting. Amer. Physiological Soc., 1990, 0193-1849/90, pp. E104–E109.

Do Benzodiazepine Ligands Contribute to Hepatic Encephalopathy?

E. Anthony Jones, Anthony S. Basile, Cihan Yurdaydin and Phil Skolnich

1. Introduction

Since 1985 (1,2) evidence has been accumulating which indicates that benzodiazepine (BZ) ligands with agonist properties may contribute to the pathogenesis of hepatic encephalopathy (HE) by potentiating the neuroinhibitory action of GABA. The status of this hypothesis in 1989 was reviewed at the International Symposium on Cirrhosis, Hepatic Encephalopathy, and Ammonium Toxicity held in Valencia (3). The purpose of this paper is to discuss developments in this field that have occurred since that review. These developments have occurred on two main fronts: (i) Detection, characterization and purification of BZ receptor ligands in tissue of animals and humans with HE; and (ii) Evaluation of the effects of BZ receptor ligands on HE in animal models and humans. The new findings are discussed after summarizing the main points covered by the review at the Valencia meeting (3).

2. Background

2.1. A Role for the $GABA_A$ Receptor Complex in HE

Major manifestations of HE, such as impaired motor function and decreased consciousness, are similar to those associated with increased GABA–mediated neurotransmission (4). Implication of the $GABA_A$/benzodiazepine receptor chloride ionophore complex (Figure 1) in the mediation of HE was originally suggested by finding that the abnormal patterns of visual evoked responses in animal models of fulminant hepatic failure (FHF) were·similar to those induced by drugs which increase GABA–ergic tone (4–7). Three other lines of evidence also suggest that GABAergic tone is increased in these models of HE. First, an increased threshold to convulsants acting by decreasing GABAergic tone is observed (4,8). Second, CNS neurons exhibit hypersensitivity to depression by agonists of the $GABA_A$ receptor complex, but not other drugs that inhibit neuronal firing (9). Finally, behavioral and electrophysiologic ameliorations of the encephalopathy can be induced by drugs that antagonize individual components of the complex (4). Increased GABAergic tone in HE does not appear to be primarily due to nonhumoral factors, such as changes in the densities or affinities of receptors on the complex (10–14) or changes in the functional status of the chloride ionophore (15). Humoral factors that could increase GABAergic tone include GABA itself and ligands for receptors on the complex which allosterically potentiate the action of

Liver Disease Section, and Laboratory of Neuroscience, NIDDK, National Institutes of Helth, Bethesda, Maryland. USA

Cirrhosis, Hyperammonemia, and Hepatic Encephalopathy,
Edited by S. Grisolia and V. Felipo, Plenum Press, New York, 1994

57

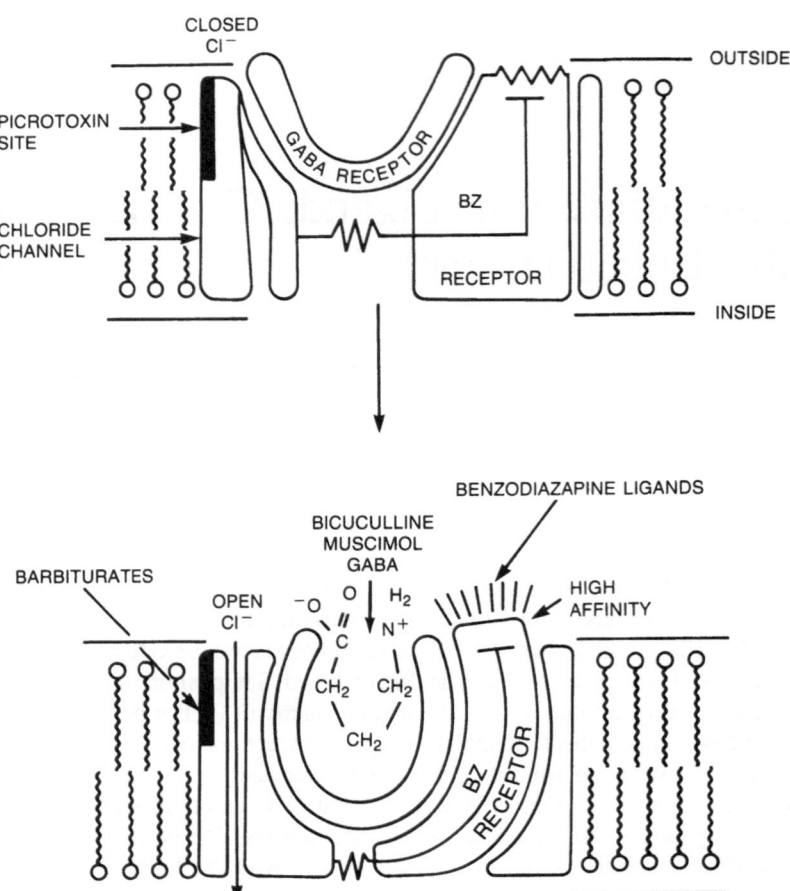

Figure 1. Diagramatic representation of the GABA$_A$/benzodiazepine receptor chloride ionophore complex in postsynaptic neural membranes in the CNS. Receptors are depicted for (i) GABA and GABA$_A$ receptor ligands e.g. muscimol, bicuculline, (ii) picrotoxin and barbiturates, and (iii) central BZ receptor ligands e.g. diazepam, flumazenil. A: shows the receptor complex in an unactivated state with the chloride channel closed. B: shows the receptor complex in the activated state with the chloride channel open. Activation is induced by GABA or a GABA agonist (e.g. muscimol) binding to GABA$_A$receptors. Activation is associated with conformational changes and opening of the chloride channel. These phenomena promote chloride conductance across the cell membrane and hyperpolarize the neuron. This mechanism is the basis of GABAergic inhibitory neurotransmission. The "gating" of the chloride channel bu GABA can be allosterically modulated by BZ receptors which exist in the form of different subtypes. BZ receptor ligands have different BZ receptor subtype specificities. (Reproduced form Biological Psychiatry 1981;16:213–229). For a more current model of the GABA$_A$ receptor complex which incorporates the configuration of subunits see reference 33.

GABA e.g. BZ agonists (16). The demonstration of increased plasma–to–brain transfer of GABA in a model of FHF (17) suggests that increased availability of GABA at GABA$_A$ receptors could contribute to increased GABAergic tone in HE. This phenomenon could occur without any increase in whole brain GABA levels (10,11).

2.2. Indirect Evidence for a Role of BZs in HE

That BZ agonists may contribute to the manifestations of HE was strongly suggested by unequivocal transient behavioral and electrophysiologic ameliorations of encephalopathy in rabbits with galactosamine (GalN)–induced FHF by the BZ antagonist flumazenil (4)

(Figure 2). This finding was shown to be neither species nor model specific (18). These observations suggest that in HE a BZ antagonist displaces endogenous ligands with agonist properties from BZ receptors, thereby reversing increased GABAergic tone by disinhibiting neurons. This inference was supported by high resolution data on the spontaneous activity (firing rate) of single Purkinje neurons in rabbit cerebellar slices. In HE BZ antagonists (flumazenil, Ro 14–7437) robustly increased the spontaneous activity of these neurons at concentrations which either did not affect or suppressed the activithy of control neurons. Furthermore, the observed increased sensitivity of these cells in HE to depression of their spontaneous activity by the GABA agonist muscimol was abolished by preincubating with a BZ antagonist (Ro 14–7437) (9). These in vitro observations can also be explained by postulating that in HE BZ antagonists displace agonists ligands from BZ receptors, thereby disinhibiting neurons. Disinhibition of neurons would lead to an increase in their spontaneous activity (excitation).

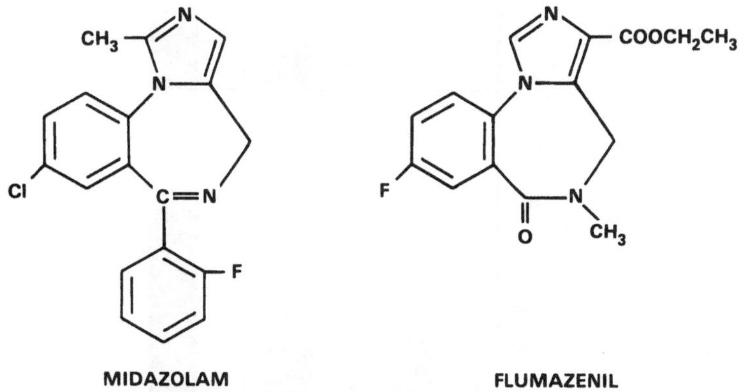

MIDAZOLAM   **FLUMAZENIL**

Figure 2. The chemical structures of midazolam and flumazenil (Ro 15–1788). Midazolam is a 1, 4–substituted BZ which acts as an agonists at a central BZ receptors. Flumazenil is an imidazobenzodiazepine which acts as an antagonist (and at high doses/concentrations, a weak partial agonist) at central BZ receptors.

3. Direct Evidence for a Role of Benzodiazepines in HE

The indirect evidence that BZs contribute to the neuronal depression of HE has been supplemented by direct evidence of their presence in models of the syndrome. The binding of radiolabeled BZ receptor ligands (^{3}H–flumazenil, ^{3}H–flunitrazepam) to unwashed brain sections from a model of HE is reduced. This phenomenon was abolished by prewashing the sections, indicating the presence in HE of ligands that reversibly bind to the BZR. In addition, muscimol further reduced radiolabeled BZ receptor binding to unwashed brain sections from the model (13). This positive (muscimol–induced) GABA shift is attributable to an increase in the affinities of the BZ receptor ligands present and indicates that they have agonist properties. In this study the distribution of radiolabeled BZ receptor ligand binding to both control brains and brains from the model of HE was heterogenous (13). These autoradiographic observations were supplemented by the demonstration in cerebrospinal fluid, brain, plasma and several peripheral organs of animal models of HE of increased activity that reversibly and competitively inhibits radiolabeled BZ receptor ligand (^{3}H–flumazenil) binding to normal brain membranes (12, 14, 19). The potency of this inhibitory activity was enhanced by GABA (12, 14), indicating the agonist nature of the BZ ligands.

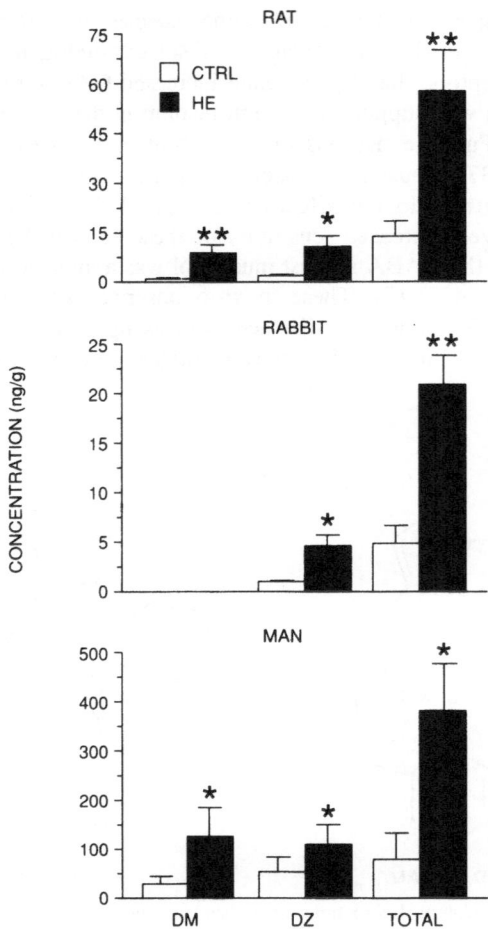

Figure 3. Levels of N–desmethyldiazepam (DM) and total BZ receptor ligand activity in the brains of rats with TAA–induced FHF, rabbits with GalN–induced FHF and humans acetaminophem–induced FHF and control brains. Concentrations are in absolute ng/g of tissue wet weight for DM and DZ an in ng/g DZ equivalents for total BZ receptor ligand activity. The absolute levels of BZ receptor ligand activity in humans with HE are 5 and 18 fold greater than those in the rat and rabbit respectively (* $p < 0.05$; **$p < 0.01$ for FHF vs controls). Reproduced from Basile et al (33).

4. Isolation and Measurement of Benzodiazepines in Animal Models of HE

Purification of whole brain extracts from models of HE and controls by HPLC revealed 3–8 peaks of inhibitory activity (i.e. activity which inhibited ^{3}H–flumazenil binding to brain membranes) with retention times similar to those of known 1,4–BZs (14, 20, 21). The chemical structure of two of these BZ receptor ligands, diazepam and N–desmethyldiazepam, was confirmed by mass spectroscopic analysis (14, 20), but the chemical nature of much of the BZ receptor ligand activity in brain in HE remains undefined. 1, 4–BZs were also found in normal brain. However, the total levels of BZ receptor ligands were significantly greater in the brains of models of HE than control brains (14, 20, 21) (Figure 3). Interestingly brain BZ levels were increased to a greater extent in rats with thioacetamide (TAA)–induced FHF than in rats with GalN–induced FHF (21). The rat the model of HE induced by TAA is considered to be superior to the rat model induced by GalN (11, 18, 22). Furthermore, brain

BZ levels were not increased in the rat with a portacaval shunt (21), which, in contrast to the TAA and GalN-induced models, does not exhibit overt liver failure and does not appear to be a satisfactory model of HE (23).

5. Isolation and Measurement of Benzodiazepines in Humans with HE

Increased BZ receptor binding activity and 1,4-BZ immunorareactivity has been demonstrated in cerebrospinal fluid, plasma and urine of patients with decompensated cirrhosis in whom there was no evidence of recent ingestion of BZs (21, 24). In cirrhotic patients the levels of BZs correlated with the severity of HE (24). In an autopsy study 6 of 11 patients who died of acetaminophen-induced FHF had brain levels of BZs that were 2-10 times higher than those in the other 5 patients or in 8 patients without liver disease who had died from cardiovascular disease or trauma. In this study brain extracts contained 4-9 chromatographic peaks of activity which inhibited ^{3}H-flumazenil binding to BZ receptors. Ultraviolet and mass spectroscopic analysis confirmed that two of the BZs present in increased concentrations in the brain of some of the patients with HE were diazepam and N-desmethyldiazepam (25). When abnormal levels of BZs were found in the brains of patients with HE the levels tended to be much greater than those found in animal models of HE (14, 20, 25) (Figure 3). The averge total concentration of BZ receptor ligands in the brains of patients with FHF was in the range that would be expected after the administration of low (anxiolytic) doses of diazepam (25). Currently, the origin of increased levels of BZ receptor ligands in liver failure is unknown.

6. Effects of Benzodiazepine Receptor Ligands on HE

6.1. Spectrum of Activities of Benzodiazepine Receptor Ligands

The BZ receptor may increase or decrease the efficiency of GABA-mediated neurotransmission depending on the nature of a BZ receptor ligand occupying the receptor. Three main classes of BZ receptor ligands are recognized: agonists, inverse agonists and antagonists. Full agonists include the classical 1,4-substituted BZs, such as diazepam. Conformational changes in the BZ receptor mediated by agonists increase GABAergic tone. Consequently agonists can induce sedation and coma. In contrast, full inverse agonists, such as the β-carbolines (e.g. methyl-6,7-dimethoxy-4-ethyl-β-carboline-3-carboxylate (DMCM)), mediate conformational changes in the BZ receptor that decrease GABAergic tone. This phenomenon can induce convulsions. Antagonists have minimal (agonist or inverse agonist) intrinsic activity and consequently their occupation of the BZ receptor alone does not alter GABA-mediated neurotransmission. However, antagonists competitively antagonize the binding of other BZ receptor ligands. Thus, antagonists tend to normalize changes in GABAergic tone induced by agonists or inverse agonists (26, 27). The spectrum of intrinsic activities of BZ receptor ligands is wide (28) (Figure 4). Many drugs classified in the central region of the spectrum have weak partial agonist or weak partial inverse agonist properties, but act predominantly as antagonists (27, 28). Thus, although flumazenil has weak partial agonist actions at high concentrations (9), it is considered to be a prototypic antagonist.

Theoretically there are two properties of BZ receptor ligands that could lead to increased neuronal activation and amelioration of manifestations of HE. One would be the ability to displace agonists from the BZ receptors. The other would be an intrinsic inverse agonist action that should result in an analeptic effect. However, the full inverse agonist, DMCM, in subconvulsive doses, was not effective in ameliorating HE in an animal model. In particular, it induced a preconvulsive state but did not efficaciously reverse the behaviorala

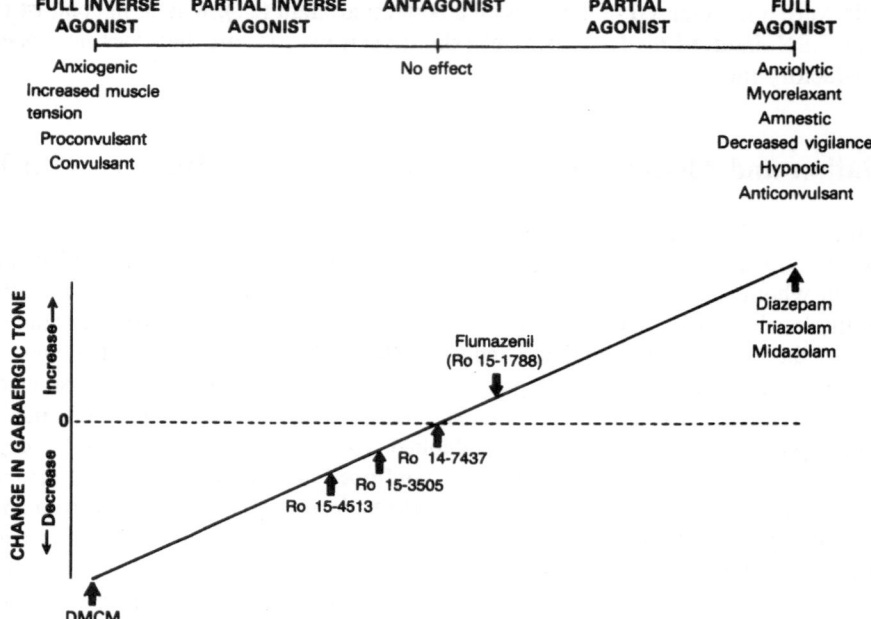

Figure 4. Spectrum of activities of central BZ receptor ligands.

manifestations of HE (18). Thus in HE a full inverse agonist is contraindicated. Conversely,full agonist, which would compound encephalopathy, is also contraindicated. However, compounds classified near the center of the spectrum of BZ receptor ligand activities (Figure 4) can subtly and incrementally modulate GABA–mediated neurotransmission through the BZ receptor (27, 28). Compounds of this type can be selected for their safety and the efficacy with which they reverse the manifestations of HE.

6.2. Rationale for Administration of BZ Antagonists in HE

A strong rationale for the administration of a BZ receptor antagonist to patients with HE is provided by the following findings in animal models of FHF: (i) BZ antagonists ameliorate behavioral and electrophysiologic manifestations of the encephalopathy (4, 18); (ii) CNS neurons are excited by BZ receptor antagonists (9); and (iii) increased levels of 1, 4–BZs are present in the brain (14, 20, 21) and CSF (19). This rationale is further justified by finding increased brain levels of 1, 4–BZs in a subpopulation of patients with FHF (25) and increased levels of BZ binding activity and BZ immunoreactivity in patients with decompensated cirrhosis (24).

6.3. Effect of Flumazenil on Human HE

Clinical application of flumazenil has been facilitated by its safety (26). As a weak partial agonist, it does not have intrinsic neuronal activating properties at moderate doses and has anticonvulsant properties at high doses (29). Its main side effect appears to be transient anxiety (30), which is probably attributable to disinhibition of neurons caused by a reduction of GABAergic tone as a result of displacement of agonist ligands from the BZ receptor (9, 26).

Anecdotal reports indicate that IV bolus injections of flumazenil induce clinical and electrophysiological (EEG or evoked response) ameliorations of HE in about 60% of patients with HE due to FHF or HE complicating cirrhosis (31–33). These ameliorations are transient, lasting 0.6 to 4hr (33), and are consistent with the rapid rate of metabolism of flumazenil (26) even in the presence of liver disease (34). Responses of HE to flumazenil are not only inconsistent, but are also typically incomplete with only some neurologic deficits being reversed and improvements in motor function being limited (31, 32). Appropriately designed controlled clinical trials are required to confirm the efficacy of flumazenil in ameliorating HE (35). In the design of such trials it is important that a history of BZ ingestion rather than a positive screening test for the presence of BZs be used as an exclusion criterion (24) and that the patients studied have sufficiently advanced stages of HE to permit a substantial flumazenil–induced effect to be detected. In addition the dose of flumazenil administered in such trials should be adequate; it has recently been estimated that a dose of about 7 mg is necessary to occupy a majority of central BZ receptors (36).

Flumazenil has been administered orally (25mg bid) to a middle–aged woman who had undergone a two–thirds partial hepatectomy and an end–to–side portacaval anastomosis simultaneously. Following the surgery she developed intractable encephalopathy that was not alleviated by comprehensive conventional therapeutic regimens. However, following the introduction of flumazenil therapy the encephalopathy rapidly subsided completely and protein tolerance became normal. Discontinuing flumazenil precipitated a recurrence of intractable encephalopathy. Restarting it was again rapidly followed by a complete and sustained remission of encephalopathy (30) (Figure 5). While this case is unusual, the syndrome of mental dysfunction that developed fulfills the criteria for chronic intractable portal–systemic encephalopathy. Accordingly it is necesary for the findings in this study to be reconciled with data obtained in studies of the effects of BZ receptor ligands on HE in animal models.

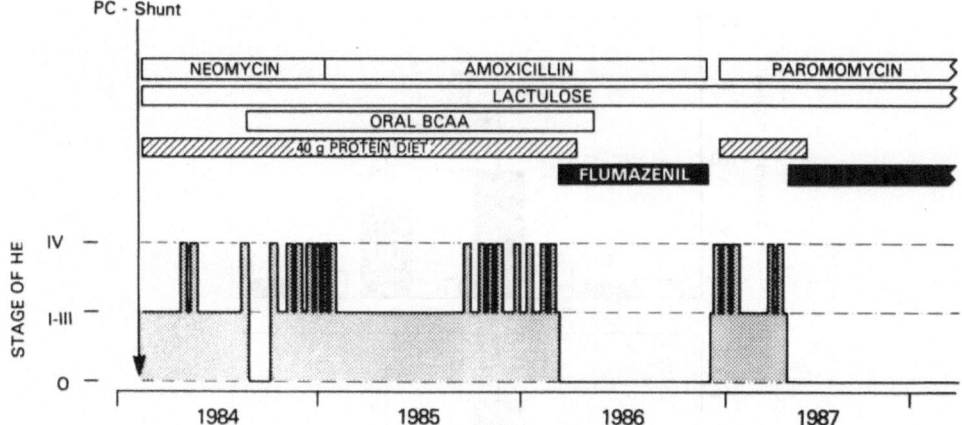

Figure 5. Remissions of chronic intractable portal systemic encephalopathy in a 42–year old woman associated the oral administration of flumazenil. Degrees of encephalopathy are depicted as = (clinically normal mental function), stages I–III and stage IV (coma). A two thirds partial hepatectomy and end–to–side portacaval shunt were followed by the development of chronic incpacitating encephalopathy with episodes of coma. Treatment with oral broad spectrum antibiotics, lactulose and dietary protein restriction did not appreciby ameliorate the encephalopathy. Addition of oral branched chain amino acid (BCAA) therapy was associated with a transient improvement in the encephalopathy. In contrasts, flumazenil administration (25 mg twice daily) was associated with complete and sustained remission of the encephalopathy which included a normalization of tolerance to dietary protein. Discontinuing flumazenil precepitated a recurrence of encephalopathy; reinstituting it was followed by a further complete and sustained remission. Re produced with modifictions from Ferenci et al (30).

6.4. Effects of Flumazenil on Models of HE

Flumazenil, in pharmacologically adequate doses, has recently been reported to induce no detectable improvement in the neurobehavioral status of rats with TAA–induced FHF (Figure 6) (37). Thus, this study failed to reproduce the modest behavioral and electrophysiologic ameliorations of HE induced by flumazenil in an earlier study using the same model (18). Flumazenil also failed to induce either behavioral or electroencephalographic improvement of encephalopathy in rats (Figure 7) (38) or rabbits (39) with acute ischemic hepatic necrosis. These observations are at variance with the behavioral and electrophysiologic ameliorations of HE induced in rabbits with GalN–induced FHF by flumazenil (4). The lack of response of encephalopathy to flumazenil in some of these models may be related to: (i) pharmacokinetic factors, including the vehicle used to dissolve flumazenil and the rapid degradation of the drug by serum esterases in the rat (40); (ii) the weak partial agonist properties of flumazenil that blunt any positive neurological responses, particulary at high doses (9); (iii) the severity of the syndrome of FHF and other model dependent factors in the models used; (iv) the adequacy of the models as models of HE, particularly acute ischemic liver injury; (v) brain levels of BZs that are not representative of the human syndrome (14, 20, 25) and (vi) BZ receptor subtype specificity. The failure of flumazenil to induce robust ameliorations of HE in animal models raises the possibility that it is not the ideal BZ receptor ligand for use in the management of HE.

6.5. Effects of other benzodiazepine receptor ligands on models of HE

The first study in which a BZ receptor ligand was applied in a model of HE was the

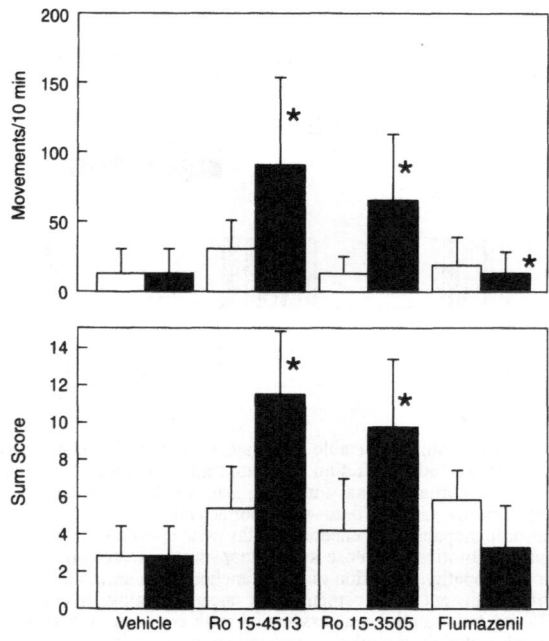

Figure 6. Motor activity (movements/18 min) and neurobehavioral testing (sum score) before and after the IP injection of vehicle, Ro 15–4513 (5 mg/Kg), Ro 15–3505 (10 mg/kg) and flumazenil (10 mg/Kg) in rats with HE due to TAA–induced FHF. Data are means ⁺ SD (N=6). *p<O.O5. Both Ro 15–4513 and Ro 15–3505 ameliorated HE whereas flumazenil was ineffective. Reproduced from Steindl et al. (37).

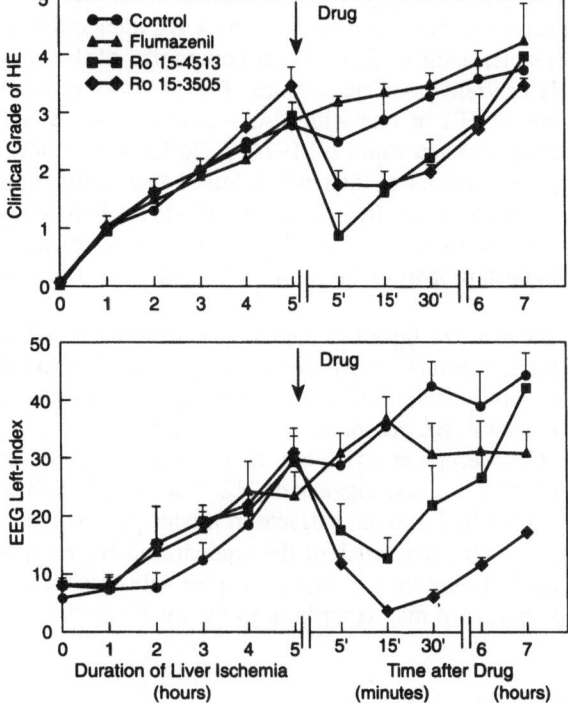

Figure 7. Effects of IV administration of flumazenil, Ro 15-3505 or Ro 15-4513 in rats with acute ischemic hepatic necrosis on the clinical grade of encephalopathy (A) and EEG left index (B). The dose of drugs was 16.6 mg/Kg. Control animals received 2 ml vehicle IV. Each line represents a group of 5 animals. Data are means ⁺SEM. Both Ro 15-3505 and Ro 15-4513 induced transient improvements in clinical grade and EEG left index, wherase flumazenil was ineffective. Reproduced from Bosman et al (38).

administration of CGS 8216 to rats with GalN-induced FHF (41). CGS 8216 is a weak partial inverse agonist (27, 28). More recently the effects of other BZ receptor ligands, on models of HE have been assessed, including Ro 15-3505, Ro 15-4513 and Ro 14-7437. These three ligands have chemical structures closely related to that of flumazenil (27, 28).

In contrast to flumazenil, Ro 15-3505 and Ro 15-4513 have partial inverse agonist properties (27, 28). Both of these ligands induce unequivocal improvements in the neurobehavioral status and electrophysiological abnormalities of rats with FHF induced by TAA or acute hepatic ischemia (Figures 6 and 7) (18, 37, 38,). The beneficial effect of these ligands on HE did not appear to be primarily due to non-specific, inverse agonist-induced analeptic effects, since the doses of these drugs that induced ameliorations of HE did not mediate any obvious behavioral or electrophysiological effects in normal animals (18, 37, 38). Furthermore, the preconvulsive state induced by DMCM was not observed after Ro 15-3505 or Ro 15-4513 (18, 37, 38). The specificity of the improved neurological state induced by these ligands for HE was further substantiated by showing that they did not ameliorate uremic encephalopathy despite the observed hyperresponsiveness of uremic animals (37).

A study with an interesting experimental design employed Ro 14-7437. First, TAA-treated rats were shown to exhibit increased sensitivity to the neurobehaviorally depressant actions of the BZ agonist flunitrazepam (41). Second, Ro 14-7437 was shown to completely reverse the behavioral effects of flunitrazepam in TAA-treated rats (42), confirming that Ro 14-7437 acts in vivo as a BZ receptor antagonist. Previously the lack of effect of this drug on the spontaneous activity of normal Purkinje neurons in vitro had been shown it to be virtually without intrinsic activity (9). The normalization of the exaggerated behavioral response of rats with acute hepatocellular necrosis by a BZ antagonist supplements findings in other studies (9, 43) which indicate that a major component of the increased sensitivity to BZs in liver failure is centrally mediated. Third, Ro 14–7437 did not induce detectable behavioral ameliorations of HE in rats with TAA–induced FHF. Finally, when rats with TAA–induced FHF were pretreated with Ro 14–7437, Ro 15–4513–induced ameliorations of HE were blocked (42). The authors of the study inferred that in this model endogenous BZ receptor ligands do not contribute to HE and that Ro 15–4513–induced ameliorations of HE are attributable to its ability to reduce GABAergic tone as a consequence of its inverse agonist rather than antagonist properties. However, other interpretations of the findings in this study are possible.

The ability of BZ receptor ligand to induce ameliorations of HE may depend on: (1) pharmacokinetics; (ii) brain levels of BZ receptor ligands; (iii) affinity for central BZ receptors; (iv) intrinsic activity; and possibly (v) BZ receptor subtype specificity (44–46). At least two subtypes of central BZ receptors are recognized–diazepam sensitive (DS) and diazepam insensitive (DI) receptors (47), but there may be many more with potential relevance to the actions of BZ receptor ligands on HE. Classical 1, 4–BZs, such as diazepam and several of the BZs identified in brain extracts in models of HE (14, 20), bind to the DS receptor. However, it is possible that much of the unidentified BZ receptor ligand activity in the brain in HE (14, 20, 25) may bind to other receptor subtypes. If such activity includes ligands with agonist properties it may contribute to the manifestations of HE.

7. Summary and Conclusions

1. Levels of BZ receptor ligands are elevated in the brain of animal models of FHF and humans with FHF. Some of these ligands have agonist properties and some are known 1, 4–BZs which bind to the DS receptor. Much of the BZ receptor ligand activity in HE is unidentified and it is possible that some may bind to receptor subtypes other than the DS receptor.

2. Average levels of BZ receptor ligands in the brain in HE do not appear to be sufficient to augment GABAergic tone to a degree that would result in severe encephalopathy (i.e. coma). However, these ligands have a heterogenous distribution in the brain and their neuroinhibitory effects may be potentiated by increased availability of GABA at GABA$_A$ receptors. Furthermore, that these ligands may contribute to HE is suggested by anecdotal reports of ameliorations of HE being induced in a majority of patients by the BZ receptor antagonist flumazenil.

3. The response of HE to flumazenil in humans is usually incomplete and in animal models may be modest. Potential explanations for these findings include pharmacokinetics, BZ receptor subtype specificity and higher levels of BZ receptor ligands in the brain in humans with HE than in animal models.

4. Certain BZ receptor ligands e.g. Ro 15–3505 and Ro 15–4513, that are structurally related to flumazenil, are more efficacious at ameliorating HE than flumazenil in animal models. These findings may be more dependent on differences in BZ receptor subtype specificity than differences in intrinsic activity. The properties of an ideal BZ receptor ligand for administration to a patient with HE would appear to be: (i) antagonist action at BZ receptors, (ii) no intrinsic activity apparent after a conventional pharmacologic dose, (iii) high specificity and affinity for BZ receptors, (iv) slow metabolism, and (v) absence of toxic effects. Promising ligands, such as Ro 15–3505, with weak partial inverse agonist actions and hence analeptic potential, require careful evaluation of their therapeutic index before clinical application.

5. BZ receptor ligands may be useful in the management of HE. Specifically, they may

be given IV: (i) to reverse effects of exogenous BZs; (ii) to aid in the differential diagnosis of encephalopathy; (iii) to provide prognostic information; and (iv) to optimize brain function. They may also be given orally with the objective of reducing dietary protein intolerance in patients with chronic liver disease.

References

1. Bransky, G., Meier, P. J., Zieger, W. H., Walser, H.,Schmid, M., and Huber, M., 1985, Reversal of hepatic /coma by benzodiazepine antagonist (Ro 15–1788), Lancet. **1**:1324–1325.
2. Basset, M. L.,Mullen K. D., Skolnick, P., and Jones, E. A., 1985, GABA and benzodiazepine receptor antagonists ameliorate hepatic encephalopathy in a rabbit model of fulminant hepatic failure, Hepatology. **5**:1032.
3. Jones, E. A., Basile, A. S., and Skolnick, P., 1990, Hepatic encephalopathy, GABA–ergic neurotransmission and benzodiazepine receptor ligands. In: Cirrhosis, Hepatic Encephalopathy, and Ammonium Toxicity. Grisolia, S., Felipo, V., Minana, M–D (Eds), Plenum Press, New York. 121–134.
4. Basset, M. L., Mullen, K. D., Skolnick, P., and Jones, E. A., 1987, Amelioration of hepatic encephalopathy by pharmacologic antagonism of the GABA$_A$–benzodiazepine receptor complex in a rabbit model of fulminant hepatic failure, Gastroenterology. **93**:1069–1077.
5. Schafer, D. F., Fowler, J. M., Brody, L. E., and Jones, E. A., 1980, Hepatic coma and inhibitory neurotransmission: The enteric bacterial flora as a source of γ–aminobutyric acid, Gastroenterology. **79**:1052.
6. Schafer D. F., Pappas, S. C., Brody, L. E., Jacobs, R., and Jones, E. A., 1984, Visual evoked potentials in a rabbit model of hepatic encephalopathy. I. Sequential changes and comparisons with drug–induced comas, Gastroenterology **86**:540–545.
7. Jones, D. B., Mullen, K. D., Roessle, M., Maynard, T., and Jones, E. A., 1987, Hepatic encephalopathy: application of visual evoked responses to test hypotheses of its pathogenesis in rats, J. Hepatol. **4**:118–126.
8. Ferreira, M. R., Gammal, S. H., and Jones, E. A., 1988, Hepatic encephalopathy: evidence of increased GABA–mediated neurotransmission in a rat model of fulminant hepatic failure, Gastroenterology. **94**:A606.
9. Basile, A. S., Gammal S. H., Mullen, K. D., Jones, E. A., and Skolnick, P., 1988, Differential responsiveness of cerebellar Purkinaje neurons to GABA and benzodiazepine receptor ligands in an animal model of hepatic encephalopathy, J. Neurosci. **8**:2414–2421.
10. Roy, S., Pomier–Layrargues, G., Butterworth, R. F., and Huet, P–M, 1988, Hepatic encephalopathy in cirrhotic and portacaval shunted dogs: Lack of changes in brain GABA uptake, brain GABA levels, brain glutamic acid decarboxylase activity and brain postsynaptic GABA receptors, Hepatology. **4**:845–849.
11. Zimmerman, C., Ferenci, P., Pifl, Ch., Yurdaydin, C., Ebner, J., Lassman, H., Roth, E., and Hortnagl, H., 1989, Hepatic encephalopathy in thioacetamide–induced acute liver failure in rats: characterization of an improved model and study of amino acid–ergic neuro-transmission, Hepatology. **9**:594–601.
12. Basile A. S., Gammal, S. H., Jones, E. A., and Skolnick, P., 1989, The GABA$_A$ receptor complex in an experimental model of hepatic encephalopathy: evidence for elevated levels of an endogenous enzodiazepine receptor ligand, J. Neurochem. **53**:1057–1063.
13. Basile, A, S., Ostrowski, N. L., Gammal, S. H., Jones, E. A., and Skolnick, P., 1990, The GABA$_A$ receptor complex in hepatic encephalopathy: autoradiographic evidence for the presence of an endogenous benzodiazepine receptor ligand, Neuropsychopharmacology. **3**:61–71.
14. Basile, A. S., 1991, The contribution of endogenous benzodiazepine receptor ligands to the pathogenesis of hepatic encephalopathy, Synapse. **7**:141–150.
15. Baker, B. L., Morrow, A. L., Vergalla, J., Paul, S. M., and Jones, E. A., 1990, Gamma-aminobutyric acid (GABA$_A$) receptor function in a rat model of hepatic encephalopathy, Metab. Brain. Dis. **5**:185–193.
16. Mullen K. D., Mendelson W. B., Martin, J. V., Roessle M., Maynard, T. F., and Jones, E. A., 1988, Could an endogenous benzodiazepine ligand contribute to hepatic encephalopathy?, Lancet. **1**:457–459.

17. Bassett M. L., Mullen K. D., Scholz, B., Fenstermacher, J. D., and Jones, E. A., 1990, Increased brain uptake of τ-aminobutyric acid in a rabbit model of hepatic encephalopathy, Gastroenterology. **98:**747-757.

18. Gammal, S. H., Basile, A. S., Geller, D., Skolnick, P., and Jones, E. A., 1990, Reversal of the behavioral and electrophysiological abnormalities of an animal model of hepatic encephalopathy by benzodiazepine receptor ligands, Hepatology. **11:**371-378.

19. Mullen, K. D., Martin, J. V., Mendelson, W. B., Kaminsky-Russ, K., and Jones, E. A., 1989, Evidence for the presence of a benzodiazepine receptor binding substance in cerebrospinal fluid of a rabbit model of hepatic encephalopathy, Metab Brain Dis. **4:**253-260.

20. Basile, A. S., Pannell, L., Jaouni, T., Gammal, S. H., Fales, H. M., Jones, E. A., and Skolnick, P., 1990, Brain concentrations of benzodiazepines are elevated in an animal model of hepatic encephalopathy, Proc Natl Acad Sci USA. **87:**5263-5267.

21. Olasmaa, M., Rothstein, J. D., Guidotti, A., Weber R. J., Paul, S. M., Spector, S., Zeneroli, M. L., Barald, M., and Costa, E., 1990, Endogenous benzodiazepine receptor ligands in human and animal hepatic encephalopathy, J. Neurochem. **55:**2015-2023.

22. Mullen, K. D., Schafer, D. F., Cuchi, P., Rossle, M., Maynard, T. F., and Jones, E. A., 1988, Evaluation of the suitability of galactosamine-induced fulminant hepatic failure as a model of hepatic encephalopathy in the rat and the rabbit. In: Advances in Ammonia Metabolism and Hepatic Encephalopathy, Soeters, P. B., Wilson, J. H. P., Meijer, A. J., Holm E., (Eds)), Elsevier, Amsterdam. 205-212.

23. Mullenm K. D., and McCullough, A. J., 1989, Problems with animal models of chronic liver disease: suggestions for improvement in standardization, Hepatology. **9:**500-503.

24. Mullen, K. D., Szauter, K, M., and Kaminsky-Russ, K., 1990, "Endogenous" benzodiazepine activity in body fluids of patients with hepatic encephalopathy, Lancet. **336:**81-83.

25. Basile, A. S., Hughes, R. H., Harrison, P. M., Murata, Y., Pannell, L., Jones, E. A., Williams, R., and Skolnick, P., 1991, Elevated brain concentrations of 1, 4-benzodiazepine in fulminant hepatic failure, N Engl J Med. **325:** 473-478.

26. Jones, E. A., Basile, A. S., Mullen, K. D., and Gammal, S. H., 1990, Flumazenil: potenial implictions for hepatic encephalopathy, Pharmacol Ther. **45:**331-343.

27. Haefely, W., Kyburz, E., Gerecke, M., and Mohler, H., 1985, Recent advances in the molecular pharmacology of benzodiazepine receptors and in the structure-activity relationships of their agonists and antagonists, Adv Drug Res. **14:**165-322.

28. Gardner, C. R., 1988, Pharmacological profiles in vivo of benzodiazepine receptor ligands, Drug Develop Res. **12:**1-28.

29. Scollo-Lavizzari, G., 1984, The anticonvulsant effect of the benzodiazepine antagonist, Ro 15-1788: an EEG study of 4 cases. Eur Neurol. **23:**1-6.

30. Ferenci, P., Grimm, G., Meryn, S., and Gangl, A., 1989, Successful long-term treatment of portal-systemic encephaopathy by the benzodiazepine antagonist flumazenil, Gastroenterology. **96:**240-243.

31. Grimm, G., Ferenci, P., Katzenschlager, R., Madl, C., Schneeweiss, B., Laggner, A. N., Lenz, K., and Gangl, A., 1988, Improvement of hepatic encephalopathy treated with flumazenil, Lancet. **2:**1392-1394.

32. Bransky, G., Meier, P. J., Riederer, E., Walser, H., Ziegler, W, H., and Schmid, M., 1989, Effects of the benzodiazepine receptor antagonist flumazenil in hepatic encephalopathy in humans, Gastroenterology. **97:**744-750.

33. Basile, A. S., Jones, E. A., and Skolnick, P., 1991, The pathogenesis and treatment of hepatic encephalopathy: evidence for the involvement of benzodiazepine receptor ligands, Pharmacol Rev. **43:**27-71.

34. Pomier-Layrarques, G., Giguere, J-F., Lavoie, J., Willems, B., and Butterworth, R. F., 1989,Pharmacokinetics of benzodiazepine antagonist Ro 15-1788 in cirrhotic patients with moderate or severe liver dysfunction, Hepatology. **10:**969-972.

35. Jones, E. A., 1991, Round table discussion on benzodiazepine antagonists. In: Progress in Hepatic Encephalopathy and Metabolic Nitrogen Exchange. Bengtsson, F., Jeppson, B., Almdal, T., Vilstrup, H., (Eds). CRC Press, Boca Raton, FL. 169-179.

36. Savic, I., Widen, L., and Stone-Elander, S., 1991, Feasibility of reversing benzodiazepine tolerance with flumazenil, Lancet. **337:**133-137.

37. Steindl, P., Püspök, A., Druml, W., and Ferenci, P., 1991, Beneficial effects of pharmacological modulation of the GABA_A-benzodiazepine receptor on hepatic encephalopathy in the rat: comparison with uremic encephalopathy, Hepatology. **14**:963–968.
38. Bosman, D. K., van der Buijs, C. A. C. G., de Haan, J. G., Maas, M. A. W., and Chamuleau, R. A. F. M., 1991, The effects of benzodiazepine receptor antagonists and partial inverse agonsits in acute hepatic encephalopathy in the rat, Gastroenterology. **101**:772–781.
39. Van der Rijt, C. C. D., de Knegt, R. J., Schalm, S. W., Terpstra, O. T., and Mechelse, K., 1990,Flumazenil does not improve hepatic encephalopathy associated with acute ischemic liver failure in the rabbit, Metab Brain Dis. **5**:131–141.
40. Mandema, J. W., Gubbens–Stribbe, J. M., and Danhof, M., 1991, Stability and pharmacokinetics of flumazenil in the rat, Psychopharmacology. **103**:384–387.
41. Baraldi, M., Zeneroli, M. L., Ventura, E., Penne, A., Pinelli, G., Ricci, P., and Santi, M., 1984, Supersensitivity of benzodiazepine receptors in hepatic encephalopathy due to fulminant hepatic failure in the rat: reversal by a benzodiazepine antagonist, Clin Sci. **67**:167–175.
42. Püspök, A., Hernath, A., Steindel, P., and Ferenci, P., 1992, Hepatic encephalopathy in rats with acute liver failure is not mediated by endogenous benzodiazepines, Gastroenterology. (In press).
43. Bakti, G., Fisch, H. V., Karlaganis, G., Minder, C., and Bircher, J., 1987, Mechanism of the excessive sedative response of cirrhotics to benzodiazepines: model experiments with triazolam, Hepatology. **7**:629–638.
44. Jones, E. A., 1991, Benzodiazepine receptor ligands and hepatic encephalopathy: Further unfolding of the GABA story, Hepatology. **14**:1286–1290.
45. Yurdaydin, C., Fromm, C., Holt, A. G., and Basile, A. S., 1991, Modulation of hepatic encephalopathy by benzodiazepine receptor ligands is selective, Hepatology. **14**:89A.
46. Yurdaydin, C., Wong, G., Basile, A. S., and Jones, E. A., 1992, Efficacy of benzodiazepine receptor antagonists in improving hepatic encepahlopathy may be related to their afinity for diazepam insensitive receptors, Hepatology. (In press).
47. Wong, G., and Skolnick, P., 1992, High affinity ligands for "diazepam–insensitive" benzodiazepine receptors, Eur. J. Pharmacol. **225**:63–68.

Effects of Hyperammonemia on Neuronal Function: NH_4^+, IPSP and Cl^--Extrusion

W. Raabe

1. Introduction

Hyperammonemia has two effects on the function of neurons: inactivation of Cl^--extrusion and depolarization of the resting membrane potential. The consequences of these two effects of hyperammonemia on the function of neurons in mammalians *in vivo* have been reviewed previously (64,65). In brief, the inactivation of Cl^--extrusion shifts the equilibrium potential of the inhibitory postsynaptic potential (IPSP, E_{IPSP}) to the level of the resting membrane potential. Under a certain condition, postsynaptic inhibition occurring at resting membrane potential, i.e., "gate-inhibition", postsynaptic inhibition cannot suppress neuronal excitation. However, when postsynaptic inhibition occurs while the neuron is temporarily depolarized, i.e., "break-inhibition", the efficacy of postsynaptic inhibition shows no significant impairment. The inactivation of Cl^--extrusion from neurons occurs without signs of changes in the effects of synaptically released GABA or glycine on neurons, and without changes in excitatory synaptic transmission. The inactivation of Cl^--extrusion and resulting dysfunction of gate-inhibition is sufficient to produce in experimental animals changes of the EEG. The brain tissue levels of NH_4^+ necessary to inactivate Cl^--extrusion and to produce encephalopathy in experimental animals are both in the same range, i.e., 0.7-1.3 mM. Therefore, it was suggested that the encephalopathy due to hyperammonemia is caused, or at least significantly contributed to, by the inactivation of Cl^--extrusion from neurons.

Evidence for the depolarization of the resting membrane potential of neurons due to hyperammonemia in mammalians *in vivo* is in part direct and in part indirect. The depolarization has several effects for the function of neurons. These effects depend on the magnitude of the depolarization. At a small depolarization presynaptic inhibition decreases and monosynaptic excitatory transmission increases. At a larger depolarization of the resting membrane potential, seizures occur and eventually all signs of synaptic transmission including seizures, cease. The depolarizing effects of NH_4^+ occurs at tissue concentrations ≥ 2 mM.

This review focusses on the earliest effect of progressively increasing hyperammonemia on neurons, i.e. a shift of E_{IPSP} to the level of the resting membrane potential due to the inactivation of Cl^--extrusion. This effect has been implicated in the initiation of the encephalopathy due to hyperammonemia (65,66,70). Although this effect of hyperammonemia is well established, the detailed mechanisms by which hyperammonemia achieves the inactivation of Cl^--extrusion and affects E_{IPSP} are not well understood.

Hyperammonemia readily increases NH_4^+ in cerebrospinal fluid (CSF) and central nervous

Departments of Neurology and Physiology, Neuroscience Graduate Program, VA Medical Center and University of Minnesota, Minneapolis, MN 55417, USA

Cirrhosis, Hyperammonemia, and Hepatic Encephalopathy,
Edited by S. Grisolia and V. Felipo, Plenum Press, New York, 1994

system (CNS) tissue (21,27,45). CNS tissue NH_4^+ increases directly and linearly with increases of blood NH_4^+ (45). The NH_4^+ concentrations referred to in this paper denote the NH_4^+–concentration the neuron is exposed to, i.e. CNS–tissue, CSF or extracellular fluid concentration of NH_4^+, unless otherwise stated.

2. NH_4^+ and IPSP

2.1. Ionic Mechanisms of IPSP

For an analysis of the effects of NH_4^+ on Cl^-–extrusion it is important to recall the fundamental observations on the ions involved in the generation of the IPSP. All the initial observations were made with intracellular recordings from neurons within the mammalian CNS. It will be expressively mentioned whenever data from *in vitro* preparations are quoted. In the mammalian CNS, E_{IPSP} depends on the salt solution used for filling the recording electrode. When recording with potassium acetate or potassium citrate filled electrodes, E_{IPSP} is more negative than the resting membrane potential. When recording with potassium chloride filled electrodes, E_{IPSP} is initially more negative than the resting membrane potential, however, E_{IPSP} changes within a few minutes to gradually become more positive than the resting membrane potential. This change of E_{IPSP} indicates a dependency of E_{IPSP} on the intracellular Cl^- concentrations because Cl^- diffuses from the electrode into the neuron recorded from. The shift of E_{IPSP} occurs without a change in the resting membrane potential (14,20,41,44,51; Raabe, W., unpublished observations).

These observations and more detailed experiments on the effects of different anions on E_{IPSP} (6,14,19,20,36) were initially interpreted as showing that the equilibrium potential for chloride ($E_{Cl}-$) is more positive than the resting membrane potential, and that E_{IPSP} is determined by the equilibrium potential for K^+ (E_K+) and $E_{Cl}-$ (18). This interpretation was based on the assumption that the ion channel opened by the inhibitory transmitter was permeable by K^+ and Cl^-, Fig. 1A. However, studies on the effects of NH_4^+ showed that E_{IPSP} could be altered without changes of a potential determined by E_K+, i.e., the afterhyperpolarization of the action potential (49,54,69). In addition, in cat trochlear motoneurons it could be shown that E_{IPSP} was more negative than E_K+, indicating that Cl^- was the dominant ion generating the IPSP (46). Thus, the involvement of K^+ in the generation of the IPSP had to be questioned. It could be inferred that in the mammalian CNS *in vivo* E_{IPSP} was more negative than the resting membrane potential (18,62) and mainly, if not exclusively determined by $E_{Cl}-$. In in vitro preparations it could be experimentally demonstrated that $E_{Cl}-$ is more negative than the resting membrane potential (16,76). Eventually, it could be conclusively shown that in mouse spinal cord neurons in tissue culture K^+ does not participate in the electrogenesis of the IPSP. The GABA and glycine activated membrane conductance was not permeable to K^+ (9). Because of the dependency of E_{IPSP} on the intracellular Cl^- concentration it was commonly presumed that $E_{Cl}-$ corresponds to E_{IPSP}, Fig. 1B. Experimentally, E_{IPSP} was used to determine $E_{Cl}-$ (cf. refs. 76,79). And, shifts of E_{IPSP} toward the resting membrane potential were thought to represent the inactivation of Cl^-–extrusion from neurons (49).

Studies on mouse spinal cord neurons in tissue culture (9) and in the crayfish stretch receptor and muscle (37–39,81) have provided new and important insights into mechanisms determining E_{IPSP}. These studies showed that the inhibitory transmitters GABA and glycine activate membrane conductances that are not only permeable to Cl^- but also to HCO_3^-. The permeability of HCO_3^- is 0.1–0.4 of that of Cl^- (9,37–39,81). The data from tissue culture experiments and the crayfish stretch receptor correlate somewhat with the observation that HCO_3^- injection into cat spinal motoneurons and cerebral cortical neurons shifts E_{IPSP} to a level more positive than the resting membrane potential (6,41). Therefore, in the mammalian

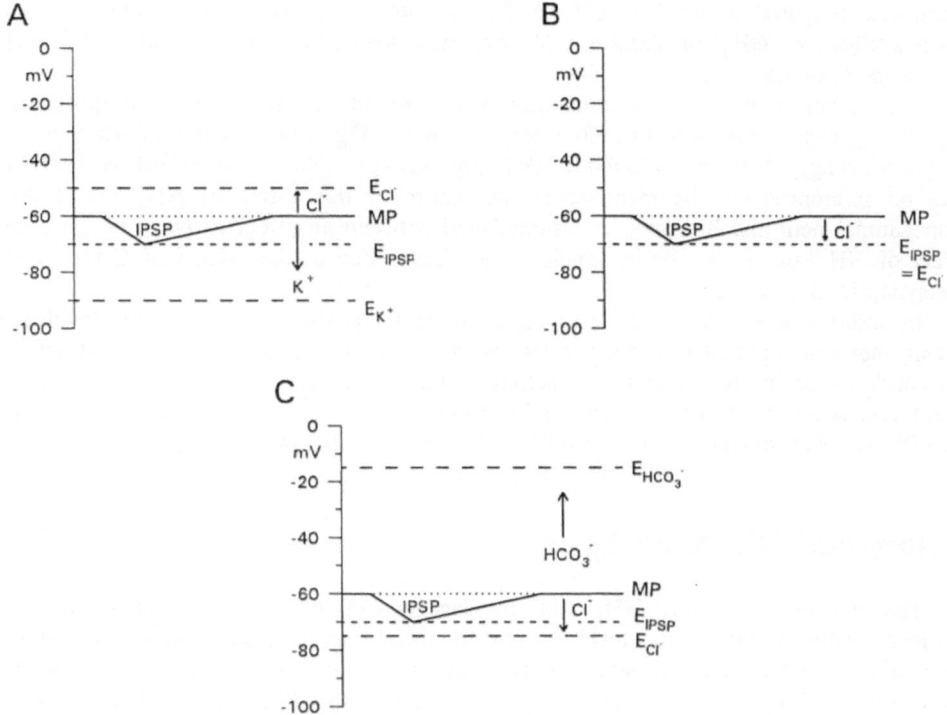

Figure 1. Models of the ionic mechanisms generating the IPSP. A– Eccles' model: IPSP is generated by influx of K^+ and efflux of Cl^-. **B** – IPSP is generated by influx of Cl^-. **C** – IPSP is generated by influx of Cl^- and efflux of HCO_3^-. MP – membrane potential.

CNS *in vivo* E_{IPSP} may not only be determined by $E_{Cl}-$ but also by the equilibrium potential for HCO_3^- ($E_{HCO3}-$). $E_{Cl}-$ is more negative than the resting membrane potential, and $E_{HCO3}-$s less negative than the resting membrane potential (5,38), Fig. 1C.

2.2. Effects of NH_4^+ and other Agents on E_{IPSP}

Intracellular, extracellular and systemic application of NH_4^+ abolishes the hyperpolarizing action of postsynaptic inhibition, i.e., shift E_{IPSP} to the level of the resting membrane potential (49,50,54). However, it is remarkable that NH_4^+ application never shifts E_{IPSP} to a level less negative than the resting membrane potential (54, Raabe, W., unpublished observations). The effect of NH_4^+ on E_{IPSP} is due to the inactivation of Cl^-–extrusion from neurons (49). These observations by Lux in cat spinal cord have been replicated in trochlear motoneurons, thalamic neurons and cerebral cortical neurons (32,47,66,67,69,71). Since glycine is the inhibitory transmitter for spinal and trochlear motoneurons, and GABA is the inhibitory transmitter in thalamus and cerebral cortex, the effect of NH_4^+ on E_{IPSP} is independent of the nature of the inhibitory transmitter.

The minimal blood plasma NH_4^+ concentration necessary to affect E_{IPSP} has been estimated as just beyond 0.1 mM. The plasma NH_4^+ concentration necessary to shift E_{IPSP} to the level of the resting membrane potential is about 1 mM. The CSF concentration necessary to show minimal effects on E_{IPSP} is ≥0.1 mM. The CSF concentration necessary to decrease E_{IPSP} by 50% is 0.23 mM. The CSF concentration necessary to maximally affect E_{IPSP} is about 0.8 mM (32). The spinal cord tissue concentration to shift E_{IPSP} to the level of the resting membrane potential is about 1 mM (68–70). In summary, plasma, CSF and CNS tissue concentrations of NH_4^+ of about 1 mM shift E_{IPSP} maximally to the level of the resting membrane potential.

It is of note that one study claims not to have found an effect of hyperammonemia on E_{IPSP} in hippocampal neurons (4). However, a review of Fig. 1 in the paper of Allen et al. (4) suggests strongly that NH_4^+ affected IPSPs in hippocampal neurons. Nevertheless, from these data no inferences can be made about the nature of this effect of NH_4^+ on IPSPs in hippocampal neurons. It cannot be distinguished between an effect of NH_4^+ on E_{IPSP} or an effect of NH_4^+ on the inhibitory conductance change due to the effects of GABA on the postsynaptic membrane.

In addition to NH_4^+, several other agents or methods shift the E_{IPSP} to the level of the resting membrane potential, but not to less negative membrane potentials, without affecting the conductance increase due to the action of the inhibitory transmitter on the neuronal membrane, see Table I. All the agents and methods listed in Table I can be expected to affect the efficacy of postsynaptic inhibition like NH_4^+ (cf. refs. 61,64,67).

3. How does NH_4^+ Affect E_{IPSP}?

The question arises how NH_4^+ and other agents shift E_{IPSP} to the level of the resting membrane potential but not to a level less negative level than the resting membrane potential. To obtain an answer to this question several issues have to be considered. (i) Experimentally it was observed that intracellular Cl^-–injection inverts the IPSP from a hyperpolarizing

Table I. Agents or methods which shift E_{IPSP} in the mammalian CNS *in vivo* to the resting membrane potential. Data from intracellular records.

Agent/Method	Application	References	Comments
	intracellular	54	
	extracellular	54	
NH_4^+	NH_4^+–salt i.v.	32,47,49,50,54,66–70	see ref. 5 for hippocampal neurons
Co^{2+}	intracellular	53,73	
Cu^{2+}	intracellular	52,53	see ref. 17 for cortical neurons
	extracellular	52	see ref. 17 for cortical neurons
Fe^{2+}	intracellular	53	
H^+	intracellular	53	
lysine	intracellular	53	
histidine	intracellular	53	
β–alanine	intracellular	53	
glycine	intracellular	23,53	
leucine	intracellular	53	
hypoglycemia		48	effect on inhibition like NH_4^+, ref. 63
Na–fluoroacetate		48	effect on inhibition like NH_4^+, ref. 63

potential to a depolarizing potential without changing the resting membrane potential (14,20,41,44,51; Raabe, W., unpublished observations). This indicates that $E_{Cl}-$ is normally more negative than the resting membrane potential, that Cl^- contributes to the electrogenesis of the membrane potential and that the permeability of the resting membrane of central mammalian neurons for Cl^- must be very low. The observations that NH_4^+ slows the recovery of E_{IPSP} from intracellular Cl^-–injections (47,49) indicate that Cl^- is not passively distributed across the neuronal membrane and that Cl^- must be extruded from neurons. The questions arise what is the nature of the mechanism of Cl^-–extrusion in neurons, and how does NH_4^+ affect this mechanism? (ii) In the mammalian CNS not only $E_{Cl}-$ but also $E_{HCO3}-$ may determine E_{IPSP}. In CO_2/HCO_3^- containing environments, as in the mammalian CNS, E_{IPSP} may be contributed to by $E_{HCO3}-$ (5,38). (iii) NH_4^+ and several other agents shift E_{IPSP} to the level of the resting membrane potential but never to a less negative level than the resting membrane potential (54; Raabe, W., unpublished observations).

3.1. Cl⁻–Extrusion

To examine membrane transport mechanisms such as Cl^-–extrusion it is desirable to change solutions on both sides of the membrane to be studied and to conduct experiments *in vivo*. In *in vivo* experiments ions and small amino acids can be added to the intracellular space with the intracellular iontophoresis technique (e.g. refs. 47,54). Larger molecules can also be added to the intracellular space with the whole cell patch clamp technique (55). One group of investigators has applied this technique even in a *in vivo* preparation (22). However, the whole cell patch clamp technique clamps the concentration gradient of Cl^- across the neuronal membrane. Therefore, this technique is not suitable to study Cl^-–extrusion. *In vivo* intracellular ion concentrations of neurons can be determined with ion–sensitive electrodes. However, this is difficult because of the small size of most neurons. The main disadvantage of *in vivo* experiments is the severely limited exchangeability of the extracellular solution. Thus, in *in vivo* experiments there is some but very limited access to the intra– and extracellular spaces. In contrast, in *in vitro* preparations there is excellent access to the extracellular space, and extracellular solutions can be easily changed. Intracellular ion concentrations can be changed by iontophoresis from sharp intracellular electrodes. Intracellular ion concentrations can be determined with ion–sensitive electrodes or ion–sensitive fluorescent indicators. Therefore, *in vitro* preparations are the preparations of choice to study membrane transport mechanisms such as Cl^-–extrusion from neurons.

Studies in many *in vitro* preparations, neuronal and other cells, have suggested that several mechanism exist which could account for extrusion of Cl^- from neurons: Na^+–dependent Cl^-/HCO_3^-–exchange, K^+/Cl^-–cotransport and transport driven by a Cl^-–ATPase (5,28). Since vertebrate neurons do not have a system for Na^+–dependent Cl^-/HCO_3^-–exchange (12,13) it is unlikely that such a mechanisms accounts for Cl^-–extrusion in the mammalian CNS (5). K^+/Cl^-–cotransport and Cl^-–transport driven by a Cl^-–ATPase have to be considered as more likely mechanisms.

3.2. K⁺/Cl⁻–Cotransport

In several *in vitro* preparations alterations of the extracellular K^+ concentration affect E_{IPSP} or the intracellular Cl^- concentration (1,2,7,42,56,57,60,76,78,79,83). An increase of extracellular K^+ shifts E_{IPSP} to the level of the resting membrane potential, and a decrease of extracellular K^+ shifts E_{IPSP} to a more negative potential. In the crayfish stretch receptor it could be demonstrated that the K^+–like ions Rb^+ and NH_4^+ (26) also decrease E_{IPSP}, whereas Cs^+ has only little effect on E_{IPSP}. The efficacy of these cations in shifting E_{IPSP} to the level of the resting membrane potential was: $Rb^+:NH_4^+:K^+:Cs^+ = 1.7:1.4:1.0:07$ (1). In addition, itwas demonstrated that known inhibitors of K^+/Cl^-–cotransport, furosemide and bumetanide,inhibit the extrusion of Cl^- from neurons (1,16,59,77,79). It was suggested that $E_{Cl}-$ is kept more negative than the resting membrane potential by a K^+/Cl^-–cotransport. It was proposed that the passive flux of K^+ from intra– to extracellular provides the energy for the extrusion of Cl^- from neurons. The K^+–site of the K^+/Cl^-–cotransport mechanism was proposed to have the selectivity sequence: $Rb^+>NH_4^+>K^+>Cs^+$ (1). The capacity of K^+/Cl^-–cotransport for Cl^- transfer across the neuronal membrane should be relatively low because frequent IPSPs or GABA application shift E_{IPSP} or E_{GABA} toward the resting membrane potential (30,57,83).

The nature of the K^+/Cl^-–cotransport mechanism is unknown. According to above discussed experimental observations, the cotransport mechanism should have sites for K^+ and Cl^-. Because of their similarity to K^+ the K^+–like modifiers of Cl^-–extrusion Rb^+, NH_4^+ and Cs^+ probably act on the K^+–site of the cotransport mechanism. The inhibition of Cl^-–extrusion by furosemide and bumetanide would suggest that the K^+/Cl^-–cotransport may be mediated by a protein analogous to capnophorin, which mediates HCO_3/Cl^-–exchange in red blood cells (28). It would be of interest to know more about the coupling of Cl^- to K^+ for purposes of cotransport. Does the coupling occur always at a fixed ratio, or is the coupling modified by the resting membrane potential, by Cl^-– and/or K^+–concentration gradients across the membrane or the intracellular Cl^- concentration? Are there other factors, e.g., intracellular pH (pH_i), which can affect K^+/Cl^-–cotransport? What is the effect of the K^+–like ion Tl^+ (26) on Cl^-–extrusion?.

How can NH_4^+ interfere with a K^+/Cl^-–cotransport driven by the transmembrane gradient of K^+? NH_4^+ has ionic properties very similar to K^+ (26). NH_4^+ may substitute for K^+ in the maintenance of the membrane potential (8). An increase of extracellular NH_4^+ may mimic in a certain way an increase of extracellular K^+ and thus decrease the driving force for a K^+/Cl^-–cotransport. However, it has to be kept in mind that in all the *in vitro* preparations, where the effects of extracellular K^+ on Cl^-–extrusion have been demonstrated, 2–5 times higher NH_4^+ concentrations than in *in vivo* preparations are necessary to inactivate Cl^-–extrusion or shift E_{IPSP} toward the level of the resting membrane potential, Table II. And, often these high NH_4^+ concentrations, e.g., 5 mM, do not shift E_{IPSP} as much to the level of the resting membrane potential as NH_4^+ 1 mM in *in vivo* preparations, cf. ref. 1 with refs. 47,54,68). Since at physiological pH a small percentage of NH_4^+ is converted to NH_3, and NH_3 readily crosses all biological membrane (15), NH_4^+ may change pH_i. A change of pH_i may then affect the inferred protein responsible for K^+/Cl^-–cotransport and decrease the coupling of Cl^- to the Cl^-–site on this protein. Although, it is expected that NH_4^+ increases pH_i, experimental studies show that NH_4^+ either has little effect on pH_i (27,45) or may even decrease pH_i (10,40).

3.3. Cl⁻–ATPase

An alternative mechanism for K^+/Cl^-–cotransport for Cl^-–extrusion from neurons is Cl^-–transport driven by a Cl^-–sensitive ATPase. Recent experimental observations on rat hippocampal neurons in tissue culture provide support for such a mechanism (31). Inoue et al. (31) monitored intracellular Cl^- with Cl^-–sensitive fluorescent dyes (11,43,80). They observed that an ATP–consuming buffer and ethacrynic acid increase intracellular Cl^-. Furosemide (0.1 mM), ouabain (0.5 mM) and an increase of extracellular K^+ from 2.5 to 10 mM had no effect on intracellular Cl^-. Furthermore, application of NH_4^+ (5 mM) plus m–chlorophenylhydrazone (CCCP) (5 μM) had also no effect on intracellular Cl^-. Inoue et al. (31) suggested that an ATP–driven pump regulates intracellular Cl^-. This correlates well with other reports by these investigators on a ethacrynic acid–sensitive and Cl^-–stimulated Mg^{2+}–ATPase in neuronal plasma membranes (24,25,33,35,72,74,75). Furthermore, hitochemical

Table II. Shift of E_{IPSP} or E_{GABA} by NH_4^+ toward the resting membrane potential.

in vivo vs. in vitro	Preparation	NH_4^+ (mM)	References
in vivo	cat spinal cord	0.8 (CSF)	32
	cat spinal cord	0.98–1.33 (tissue)	68–70
in vitro	hippocampal slice, CA1	2	29
	hippocampal slice, CA1	4	3
	hippocampal slice, CA3	2.5	76
	hippocampal neurons in tissue culture	5	82
	crayfish stretch receptor	5	58
	crayfish stretch receptor	5	16
	crayfish stretch receptor	5–20	1
	crayfish stretch receptor	5	2
	frog spinal cord	2	61

stains demonstrated the presence of such Cl^-–stimulated Mg^{2+}–ATPase in rat spinal motoneurons *in vivo* (34).

The lack of an effect of NH_4^+ on the ATP–driven Cl^-–pump/Cl^-–stimulated ATPase is surprising. This cannot be explained by the simultaneous application of NH_4^+ and the protonophore CCCP. Although NH_4^+ can be expected to increase pH_i, this is not necessarily so. As already mentioned above, there is indirect evidence that NH_4^+ causes no change in pH_i. The expected increase in pH_i does not take place because NH_4^+ stimulates glycolysis and increases lactate production (27,45). Moreover, there is direct evidence that NH_4^+ may actually decrease pH_i in neuronal tissue (10,41). CCCP can be expected to decrease pH_i. Instead of preventing with NH_4^+ the decrease of pH_i caused by CCCP, the simultaneous application of NH_4^+ and CCCP by Inoue et al. (31) may have resulted in a rather large decrease of pH_i. This would mimic then the effect of the ATP–consuming buffer used by Inoue et al. (31) This buffer, which consisted of 2–deoxyglucose, hexokinase and CCCP, can be expected to decrease pH_i. However, the ATP–consuming buffer increased intracellular Cl^- whereas the application of NH_4^+ plus CCCP did not.

The suggested mechanisms for Cl^-–extrusion from neurons, K^+/Cl^-–cotransport and a Cl^-–stimulated ATPase, are quite divergent. The K^+/Cl^-–cotransport system is sensitive to increases of extracellular K^+, furosemide and NH_4^+ whereas the Cl^-–stimulated ATPase is not. One possible but unlikely cause for the different mechanisms proposed for Cl^-–extrusion are the different techniques used to measure intracellular Cl^-. All the studies which support the existence of a K^+/Cl^-–cotransport used intracellular recordings. Intracellular Cl^- was always measured as E_{IPSP} or E_{GABA} and then calculated from the Nernst equation or Goldman–Hodgkin–Katz equation. In contrast, the main study supporting a Cl^-–stimulated ATPase measured Cl^- directly with a Cl^-–sensitive fluorescent indicator. Further differences between the postulated mechanisms for Cl^-–extrusion are summarized in Table III.

3.4. E_{IPSP} and HCO_3^-

The studies of Bormann (9), Voipio et al. (81) and Kaila et al. (37–39) suggest that the conductances activated in the postsynaptic membrane by GABA and glycine are permeable to HCO_3^-. The equilibrium potential for HCO_3^- is less negative than the resting membrane

Table III. Effects of various agents on K^+/Cl^-–cotransport and ATP-driven Cl^-–pump.

Agent	K^+/Cl^-–cotransport	ATP-driven Cl^-–pump
extracellular K^+ 10 mM	decrease (1,76,77,79)	no effect (31)
furosemide 0.1 mM 0.5 –2.0 mM	no data decrease (1,59,76,79)	no effect (31) no data
bumetanide	decrease (77)	no data
ouabain	0.1 mM, decrease (1)	0.5 mM, no effect (31)
NH_4^+ ≥5 mM	decrease (1,2,16)	no data
NH_4^+ + CCCP	no data	no effect (31)
ATP–consuming buffer	no data	decrease (31)
ethacrynic acid	no data	decrease (31)

potential (38). Since $E_{HCO3^-}=E_{H^+}$ and $[H^+]_i x[HCO_3^-]_i=[H^+]_o x[HCO_3^-]_o$ (5,38), E_{IPSP} should become more positive with an increase in pH_i and/or decrease in extracellular HCO_3^-. Thus, mechanisms affecting pH_i and/or extracellular HCO_3^- may affect E_{IPSP}. Since theoretically NH_4^+ increases pH_i, particularly high NH_4^+ concentrations such as those necessary to affect E_{IPSP} in *in vitro* preparations, 2–5 mM, the shift of E_{IPSP} toward the resting membrane potential by NH_4^+ could possibly be explained as a sequelae of an increase of pH_i. However, for two reasons this mechanism of action of NH_4^+ is unlikely. *In vivo* only 1 mM of NH_4^+ is necessary to shift E_{IPSP} to the level of the resting membrane potential. This small NH_4^+ concentration is unlikely to produce any significant increase of pH_i (38). And, as mentioned already above, in *in vivo* experiments NH_4^+ does not change pH_i (27,45). In the cerebral cortical slice *in vitro* NH_4^+ may even decrease pH_i (10,40). Thus, it is most unlikely that NH_4^+ shifts E_{IPSP} toward the resting membrane potential by shifting E_{HCO3^-} to a more positive level.

There is also a third reason which makes an action of NH_4^+ on E_{IPSP} via a depolarizing shift of E_{HCO3^-} unlikely. *In vivo* as well as in *vitro* NH_4^+ shifts E_{IPSP} only towards the level of the resting membrane potential but never beyond the resting membrane potential. If NH_4^+ would affect E_{IPSP} via a depolarizing shift of E_{HCO3^-}, it would be expected that NH_4^+ can change E_{IPSP} from a more negative level than the resting membrane potential to a more positive level than the resting membrane potential (cf. ref. 6).

4. Conclusions

An effect of NH_4^+ on postsynaptic inhibition, the shift of E_{IPSP} toward the resting membrane potential, which is due to the inactivation of Cl^-–extrusion from neurons, has been proposed to cause or contribute significantly to the development of the encephalopathy due to ammonia intoxication. Therefore, it is desirable to develop agents or methods which may prevent the effects of NH_4^+ on E_{IPSP} and Cl^-–extrusion. However, to develop these agents or methods it is necessary to understand the electrogenesis of the IPSP and the mechanisms of action of NH_4^+ on the IPSP. This review outlined that at present the electrogenesis of the

IPSP in neurons of the mammalian CNS *in vivo* is only partly understood. Although it is well substantiated that Cl^- is the major ion generating the IPSP, the participation of other ions like HCO_3^- has to be explored in further detail. The mechanism which keeps the intracellular concentration of Cl^- lower than expected from passive distribution, i.e. Cl^-–extrusion, is controversial. A K^+/Cl^-–cotransport and an ATP–driven pump/Cl^-–stimulated ATPase have been suggested. However, drugs or agents which affect one transport mechanism do not affect the other. These discrepancies have to be clarified by combining intracellular recordings from neurons with imaging intracellular Cl^- and H^+–concentrations by means of Cl^- and H^+– sensitive fluorescent dyes. Another issue to be clarified is the requirement in *in vitro* preparations for much higher NH_4^+ concentrations to shift E_{IPSP} to the level of the resting membrane potential than in *in vivo* preparations. Only when the electrogenesis of the IPSP, the transmembranal transport mechanisms for Cl^- and possibly other ions as well as the effects of NH_4^+ on these mechanisms are fully understood, agents or methods can be designed to interfere with the effects of NH_4^+ on E_{IPSP} in order to prevent or ameliorate the encephalopathy due to hyperammonemia.

Acknowledgement. This work was supported by a research grant from the department of Veterans Affairs.

References

1. Aickin, C. C., Deisz, R. A., and Lux, H. D., 1982, Ammonium action on post– synaptic inhibition in crayfish neurones: implications for the mechanism of chloride extrusion, J. Physiol. (Lond.) **329**:319–339.
2. Aickin, C. C., Deisz, R. A., and Lux, H. D., 1984, Mechanism of chloride transport in crayfish stretch receptor neurones and guinea pig vas deferens: implications for inhibition mediated by GABA, Neurosci. Lett. **47**:239–244.
3. Alger, B. E., and Nicoll, R. A., 1983, Ammonia does not selectively block IPSPs in rat hippocampal pyramidal cells, J. Neurophysiol. **49**:1381–1391.
4. Allen, G. I., Eccles, J. C., Nicoll, R. A., Oshima, T., and Rubia, F. J., 1977, The ionic mechanism concerned in generating the i.p.s.ps of hippocampal pyramidal cells, Proc. R. Soc. Lond. B. **198**:363–384.
5. Alvarez–Leefmans, F. J., 1990, Intracellular Cl^- regulation and synaptic inhibition in vertebrate and invertebrate neurons. In: Chloride Channels and Carriers in Nerve, Muscle and Glial Cells, F. J., Alvarez–Leefmans and J. M., Russell (Eds.), Plenum Press, New York. pp.109–158.
6. Araki, T., Ito, M., and Oscarsson, O., 1961, Anion permeability of the synaptic and non–synaptic motoneurone membrane, J. Physiol. (Lond.) **159**:410–435.
7. Ascher, P., Kunze, D., and Neild, T. O., 1976, Chloride distribution in Aplysia neurones, J. Physiol. (Lond.) **256**:441–464.
8. Binstock, L., and Lecar, H., 1969, Ammonium currents in the squid giant axon, J. Gen. Physiol. **53**:342–361.
9. Bormann, J., Hamill, O. P., and Sakmann, B., 1987, Mechanism of anion permeation through channels gated by glycine and γ–aminobutyric acid in mouse cultured spinal neurones, J. Physiol. (Lond.) **385**:243–286.
10. Brooks, K. J., Kauppinen, R. A., Williams, S. R., Bachelard, H. S., Bates, T. E., and Gadian, D. G., 1989, Ammonia causes a drop in intracellular pH in metabolizing cortical brain slices. A [^{31}P]– and [^{1}H]nuclear magnetic resonance study, Neuroscience. **33**:185–192.
11. Chao, A. C., Dix, J. A., Sellers, M. C., and Verkman, A. S., 1989, Fluorescence measurement of chloride transport in monolayer cultured cell, Biophys, J. **56**: 1071–1081.
12. Chesler, M., 1986, Regulation of intracellular pH in reticulospinal neurones of the lamprey, Petromyzon marinus, J. Physiol. (Lond.) **381**:241–261.
13. Chesler, M., 1987, pH regulation in the vertebrate central nervous system: Microelectrode studies in the brain stem of the lamprey, Can, J. Physiol. Pharmacol. **65**:986–993.

14. Coombs, J. S., Eccles, J. C., and Fatt, P., 1955, The specific ionic conductances and the ionic movements across the motoneuronal membrane that produce the inhibitory–post synaptic potential, J. Physiol. (Lond.) **130**:326–373.
15. Cooper, A. J. L., and Plum, F., 1987, Biochemistry and physiology of brain ammonia, Physiol. Rev. **67**:440–519.
16. Deisz, R. A., and Lux, H. D., 1882, The role of intracellular chloride in hyperpolarizing post–synaptic inhibition of crayfish stretch receptor neurones, J. Physiol. (Lond.) **326**:123–138.
17. Dreifuss, J. J., Kelly, J. S., and Krnjevic, K., 1969, Effects of copper on cortical neurones, Brain Res. **13**:607–611.
18. Eccles, J. C., 1964, The Physiology of Synapses, Springer, New York. pp. 173–188.
19. Eccles, J. C., Eccles, R. M., and Ito, M., 1964, Effects produced on inhibitory postsynaptic potentials by the coupled injection of cations and anions into motoneurons, Proc. R. Soc. Lond. B **160**:197–210.
20. Eccles, J. C., Nicoll, R. A., Oshima, T., and Rubia, F. J., 1977, The anionic permeability of the inhibitory postsynaptic membrane of hippocampal pyramidal cells, Proc. R. Soc. Lond. **198**:345–361.
21. Ehrlich, M., Plum, F., and Duffy, T. E., 1980, Blood and brain ammonia concentrations after portacaval anastomosis, Effects of acute ammonia loading, J. Neurochem. **34**:1538–1542.
22. Ferster, D., and Jagadeesh, B., 1992, EPSP-IPSP interactions in cat visual cortex studied with *in vivo* whole–cell patch recording, J. Neurosci. **12**:1262–1274.
23. Globus, A., Lux, H. D., and Schubert, P., 1968, Somadendritic spread of intracellularly injected tritiated glycine in cat spinal motoneurons, Brain Res. **11**:440–445.
24. Hara, M., Miwa, S., Fujiwara, M., and Inagaki, C., 1982, Effects of several anions on ethacrynic acid high– and low–sensitive Mg–ATPase activities in microsomal fractions from rabbit cortical gray matter, Biochem. Pharmacol. **31**:877–879.
25. Hara, M., Fujiwara, M., and Inagaki, C., 1982, Non–mitochondrial origin of thacrynic acid high–sensitive Mg^{2+}–ATPase activity in microsomal fractions from rabbit cortical gray matter, Biochem. Pharmacol. **31**:4077–4079.
26. Hille, B., 1992, Ionic Channels of Excitable Membranes, Sinauer Associates, underland, MA, p. 268–276, 1992.
27. Hindtfeldt, B., and Siesjö, B. K., 1971, Cerebral effects of acute ammonia intoxication, I, The influence on intracellular and extracellular acid–base parameters, Scand, J. clin. Lab. Invest. **28**:353–364.
28. Hoffmann, E. K., 1986, Anion transport systems in the plasma membrane of vertebrate cells, Biochem. Biophys. Acta **864**:1–31.
29. Hotson, J. R., and Prince, D. A., A calcium–activated hyperpolarization follows repetitive firing in hippocampal neurons, J. Neurophysiol. **43**:409–419.
30. Huguenard, J. R., and Alger, B. E., 1986, Whole–cell voltage clamp study of the fading of GABA–activated currents in acutely dissociated hippocampal neurons, J. Neurophysiol. **56**:1–18.
31. Inoue, M., Hara, M., Zeng, X.–T., Hirose, T., Onishi, S., Yasikura, T., Uriu, T., Omori, K., Minato, A., and Inagaki, C., 1991, An ATP–driven Cl^- pump regulates Cl^- concentration in rat hippocampal neurons, Neurosci. Lett. **134**:75–78.
32. Iles, J. F., and Jack, J. J. B., 1980, Ammonia: assessment of its action on postsynaptic inhibition as a cause of convulsions, Brain **103**:555–578.
33. Inagaki, C., Tanaka, T., Hara, M., and Shiko, J., 1985, Novel microsomal anion– sensitive Mg^{2+}–ATPase in rat brain, Biochem. Pharmacol. **34**:1705–1712.
34. Inagaki, C, Oda, W., Kondo, K., and Kusumi, M., 1987, Histochemical demonstration of Cl^-–ATPase in rat spinal motoneurons, Brain Res. 375–378.
35. Inagaki, C., and Shiroya, T., 1988, ATP-dependent Cl^- uptake by plasma membrane vesicles from rat brain, Biochem. Biophys. Res. Comm. **154**:108– 112.
36. Ito, M., Kostyuk, P. G., and Oshima, T., 1962, Further study on anion permeability in cat spinal motoneurones, J. Physiol. (Lond.) **164**:150–156.
37. Kaila, K., and Voipio, J., 1987, Postsynaptic fall in intracellular pH induced by ABA–activated bicarbonate conductance, Nature. **330**:163–165.
38. Kaila, K., and Voipio, J., 1990, GABA-activated bicarbonate Conductance. In F.J. lvarez–

Leefmans and J.M. Russell (Eds.) Chloride Channels and Carriers in Nerve, Muscle, and Glial Cells, Plenum Press, New York. pp. 331–352.

39. Kaila, K., Pasternack, M., Saarikoski, J., and Voipio, J., 1989, Influence of GABA- gated bicarbonate conductance on potential, current and intracellular chloride in crayfish muscle fibers. J. Physiol. (Lond.) **416:**161–181.

40. Kauppinen, R. A., Williams, S. R., Brooks, K. J., and Bachelard, H. S., Effects of ammonium on energy metabolism and intracellular pH in guinea pig cerebral cortex studied by ^{31}P and ^{1}H nuclear magnetic resonance spectroscopy, Neurochem. Int. **19:**495–504.

41. Kelly, J. S., Krnjevic, K., Morris, M. E., and Yim, G. K. W., Anionic permeability of cortical neurones, Exp. Brain Res. **7:**11–34.

42. Korn, S. J., Giacchino, J. L., Chamberlin, N. L., and Dingledine, R., 1987, Epileptiform burst activity induced by potassium in the hippocampus and its regulation by GABA–mediated inhibition, J. Neurophysiol. **57:**325–340.

43. Krapf, R., Berry, C. A., and Verkman, A. S., 1988, Estimation of intracellular chloride activity in isolated perfused rabbit proximal tubules using a fluorescent probe, Biophys, J. **53:**955–962.

44. Krnjevic, K., and Schwartz, S., 1967, The action of γ–aminobutyric acid on cortical neurones, Exp, Brain Res. **3:**320–336.

45. Lin, S., and Raabe, W., 1985, Ammonia intoxication: effects on cerebral cortex and spinal cor d, J. Neurochem. **44:**1252–1258.

46. Llinas, R., and Baker, R., 1972, A chloride–dependent inhibitory postsynaptic potential in cat trochlear motoneurons, J. Neurophysiol. **35:**484–492.

47. Llinas, R., Baker, R., and Precht, W., 1974, Blockade of inhibition by ammonium acetate action on chloride pump in cat trochlear motoneurons, J. Neurophysiol. **37:**522–533.

48. Loracher, C., and Lux, H. D., 1974, Impaired hyperpolarising inhibition during insulin hypoglycemia and fluoroacetate poisoning, Brain Res. **69:**164–169.

49. Lux, H. D., 1971, Ammonium and chloride extrusion: hyperpolarizing inhibition in cat spinal motoneurons, Science. **173:**555–557.

50. Lux, H. D., 1974, Fast recording ion specific microelectrodes: their use in pharmacological studies in the CNS, Neuropharmacology. **13:**509–517.

51. Lux, H. D., and Klee, M. R., 1962, Intracelluläre Untersuchungen über den Einfluß hemmender Potentiale im motorischen Cortex. Die Wirkung elektrischer Reizung unspezifischer Thalamuskerne, Arch, Psychiat, Nervenkr. **203:**648– 666.

52. Lux, H. D., and Globus, A., 1968, Effect on IPSPs of cat spinal motoneurones due to intra– and extracellular iontophoresis of $CuSO_4$ Brain Res. **9:**377–380.

53. Lux, H. D., and Schubert, P., 1969, Postsynaptic inhibition: Intracellular effects of various ions in spinal motoneurons, Science. **166:**625–626.

54. Lux, H. D., Loracher, C., and Neher, E., 1970, The action of ammonium ions on postsynaptic inhibition in cat spinal motoneurons, Exp, Brain Res. **11:**431–447.

55. Marty, A., and Neher, E., 1983, Tight–seal whole cell recording. In: Single– hannel Recording,. B. Sakmann and E. Neher (Eds.), Plenum, New York, pp. 107–122.

56. Matthews, G., and Wickelgren, W. O., 1979, Glycine, GABA, and synaptic inhibition of reticulospinal neurones of lamprey, J. Physiol. (Lond.) **293:**393– 415.

57. McCarren, M., and Alger, B. E., 1985, Use–dependent depression of IPSPs in rat hippocampal pyramidal cells in vitro, J. Neurophysiol. **53:**557–571.

58. Meyer, H., and Lux, H. D., 1974, Action of ammonium on a chloride pump, flügers Arch. **350:**185–195.

59. Misgeld, U., Deisz, R. A., Dodt, H. U., and Lux, H. D., The role of chloride transport in postsynaptic inhibition of hippocampal neurons, Science. **232:**1413–1415.

60. Motokizawa, F., Reuben, J. P., and Grundfest, H., Ionic permeability of the inhibitory postsynaptic membrane of lobster muscle fibers, J. gen. Physiol. **54:**437–461.

61. Nicoll, R. A., The blockade of GABA mediated responses in the frog spinal cord by ammonium ions and furosemide, J. Physiol. (Lond.) **283:**121–132.

62. Raabe, W., and Lux, H. D., Hyperpolarizing inhibition in cat motor cortex, Brain Res. **52:**389– 393.

63. Raabe, W., 1981, Ammonia and disinhibition in cat motor cortex by ammonium acetate, monofluoroacetate and insulin–induced hypoglycemia, Brain Res. **210:**311–322.

64. Raabe, W., 1989, Neurophysiology of ammonia intoxication. In R. Butterworth and G. Pomier-Layrargues (Eds.), Hepatic Encephalopathy: Pathophysiology and Treatment. Humana Press, Clifton, N. J., pp. 49–77.

65. Raabe, W., 1990, Effects of NH_4^+ on the function of the CNS. In S. Grisolia et al. (Eds.), Cirrhosis, Hepatic Encephalopathy and Ammonium Toxicity, Plenum Press, New York, pp. 99–120.

66. Raabe, W., And Lux, H. D., 1972, Studies on extracellular potentials generated by synaptic activity on single cat motor cortex neurons. In H. Petsche and M.A.B. Brazier (Eds.), synchronization of EEG Activity in Epilepsies, Springer, New York, pp.46–58.

67. Raabe, W., and Gumnit, R. J., 1975, Disinhibition in cat motor cortex by ammonia. J. Neurophysiol. **38**:347–355.

68. Raabe, W., and Lin, S., 1983, Ammonia intoxication and hyperpolarizing postsynaptic inhibition, Exp. Neurol. **82**:711–715.

69. Raabe, W., and Lin, S., Ammonia, postsynaptic inhibition and CNS–energy state, Brain Res. **303**:67–76.

70. Raabe, W., and Lin, S., 1985, Pathophysiology of ammonia intoxication, Exp. Neurol. **87**:519–532.

71. Singer, W., and Creutzfeldt, O., 1974, Unpublished observations quoted by reutzfeldt, O. and ouchin, J. in Neuronal Basis of EEG Waves, In: Handbook of Electroencephalography and Clinical Neurophysiology, A. Remond (Ed.), Elsevier, Amsterdam, Vol. 2, Part C, pp. 2C3–2C55.

72. Shiroya, T., Fukunaga, R., Akashi, K., Shimada, N., Takagi, Y., NIshino, T., Hara, M., and Inagaki, C., 1989, An ATP–driven Cl^- pump in the brain, J. Biol. Chem. **264**:17416–16421.

73. Sypert, G. W., and Bidgood, W. D., 1977, Effect of intracellular cobalt ions on postsynaptic inhibition in cat spinal motoneurons, Brain Res. **134**:372–376.

74. Tanaka, T., Inagaki, C., Matsuda, K., and Takaori, S., 1986, Characteristics of thacrynic acid highly sensitive Mg^{2+}–ATPase in microsomal fractions of the rat brain: functional molecular size, inhibition by SITS and stimulation by Cl^-. Jpn. J. Pharmacol. **42**:351–359.

75. Tanaka, T., Inagaki, C., Kunigi, Y., and Takaori, S., Solubilization and separation of ethacrynic acid (EA) highly sensitive and EA less sensitive Mg^{2+}–ATPase in the rat brain. Jpn. J. Pharmacol. **43**:205–212.

76. Thompson, S. M., Deisz, R. A., and Prince, D. A., 1988, Relative contributions of passive equilibrium and active transport to the distribution of chloride in mammalian cortical neurons, J. Neurophysiol. **60**:105–124.

77. Thompson, S. M., Deisz, R. A., and Prince, D. A., 1988, Outward chloride/cation co– transport in mammalian cortical neurons, Neurosci. Lett. **89**:49–54.

78. Thompson, S. M., and Gähwiler, B. H., Activity–dependent disinhibition. I.Repetitive stimulation reduces IPSP driving force and conductance in the hippocampus in vitro, J. Neurophysiol. **61**:501–511.

79. Thompson, S. M., and Gähwiler, B. H., Activity–dependent disinhibition. II. Effects of extracellular potassium, furosemide, and membrane potential on E_{Cl}– in hippocampal CA3 neurons, J. Neurophysiol. **61**:512–523.

80. Verkman, A. S., Sellers, M. C., Chao, A. C., Leung, T., and Ketcham, R., Synthesis and characterization of improved chloride–sensitive fluorescent indicators for biological application, Anal. Biochem. **178**:355–361.

81. Voipio, J., Rydqvist, B., and Kaila, K., The reversal potential of GABA–activated current (E_{GABA}) may be sensitive to metabolic production of CA_2/HCO_3. Eur. J. Neurosci. ENA–Suppl. p. 61 (abstract).

82. Walton, M. K., Raabe, W., and Barker, J. L., 1987, Effect of ammonium ion on E_{GABA} in cultured rat hippocampal neurons, Soc. Neurosci. Abstr. **13**(2):957.

83. Wong, R. K. S., and Watkins, D. J., 1982, Cellular factors influencing GABA response in hippocampal pyramidal cells, J. Neurophysiol. **48**:938–951.

Activation of NMDA Receptor Mediates the Toxicity of Ammonia and the Effects of Ammonia on the Microtubule–Associated Protein MAP–2.

Vicente Felipo, Eugenio Grau, Maria–Dolores Miñana and Santiago Grisolia.

1. Introduction

Hepatic encephalopathy is one of the main causes of death in western countries. In Spain, nearly 12,000 people die every year by liver cirrhosis (≈8,000) or other hepatic failures; this represents approximately 4% of the total number of deaths. In spite of much work, the actual cause of hepatic coma and death remains unclear.

Several hypothesis have been proposed to explain the pathogenesis of hepatic encephalopathy, including depletion of metabolites required for the formation of metabolic energy in brain (1); neural inhibition mediated by the formation of "false neurotransmitters" (2) as a consequence of plasma amino acid imbalance and of increased activity of the blood-brain neutral amino acid transport system (3); synergism between mercaptans and ammonia or fatty acids in the production of coma (4); and activation of the γ–aminobutyric acid (GABA) inhibitory neurotransmitter system (5). However, none of these hypothesis has been completely supported by the experimental observations.

Hyperammonemia is considered one of the most important factors responsible for the mediation of hepatic encephalopathy. Elevated levels of blood ammonia accompany a number of human liver diseases, including cirrhosis, acute liver failure, Reye's syndrome and hereditary deficiencies of the urea cycle enzymes. Most of the above hypothesis involve hyperammonemia as one of the possible causes of hepatic encephalopathy. Also, the classical clinical treatments of hepatic encephalopathy are directed towards decreasing the ammonia levels in blood, by decreasing its formation by the intestinal flora (using antibiotics) and/or reducing its transport into the bloodstream by colonic acidification (6). However, it has not been definitively demonstrated whether hyperammonemia is or not the factor responsible for the pathogenesis of hepatic encephalopathy. Ammonia toxicity was first reported a century ago (7) by Pavlov and his group. They excluded the liver form the circulation and found that when dogs treated in this way were fed meat, they developed hyperammonemia, which was associated with coma and led to the death of the animals. The concentration of ammonia is maintained at very low levels in most animal tissues. A five– to ten–fold increase of the normal ammonia concentrations in blood induces toxic effects in most animal species, with

Instituto de Investigaciones Citológicas de la Fundación Valenciana de Investigaciones Biomédicas. Amadeo de Saboya, 4. 46010 Valencia. Spain.

Cirrhosis, Hyperammonemia, and Hepatic Encephalopathy,
Edited by S. Grisolia and V. Felipo, Plenum Press, New York, 1994

functional disturbances of the central nervous system. Acute injection of high doses of ammonia lead to the rapid death of the animal. However, the molecular mechanism of acute ammonia toxicity has not been clarified. Also the possible contribution of mild, sustained hyperammonemia to the neurologic disturbances of hepatic encephalopathy remains unclear.

To study these questions it is important to have suitable animal models of hyperammonemia and of hepatic encephalopathy. Several models have been described in the literature. The first model used was the Eck's fistula or portacaval anastomosis in dogs (8) and later in rats (9, 10). Other models are based on the use of hepatotoxic compounds such as galactosamine (11), acetaminophen (e.g. 12), thioacetamide or CCl_4. These models are intended to reproduce the effects of chronic or fulminant hepatic failure and produce in the animals, in addition to hyperammonemia, a complex syndrome which usually includes liver atrophy, weight loss and muscle wasting. Therefore, these models are not suitable to study the effects of "pure" hyperammonemia nor to clearly discern its contribution to the neurologic disfunction found in hepatic encephalopathy. There are also models to study the effects of "pure" hyperammonemia, without impairment of liver function. These models include continuous infusion of ammonia (e.g. 13) and injection of urease to produce ammonia from the endogenous urea (14). Although these are good models of hyperammonemia, they are not suitable for long term studies. Another model of hyperammonemia, useful for long term studies, with no impairment of liver function and consisting of feeding rats an ammonium containing diet has also been used (15–16).

2. Effects of Hyperammonemia on Brain Microtubules

Using this model, we studied the effect of hyperammonemia on brain protein synthesis and found that chronic, mild hyperammonemia induces the synthesis and accumulation of

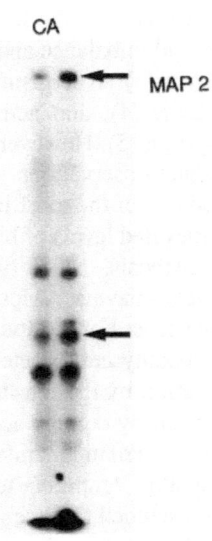

Figure 1. Patterns of proteins phosphorylated in microtubules from control and hyperammonemic rats. Rats were fed control or ammonium diet for 15 days and microtubules were isolated from brain. Protein phosphorylation was assayed in a volume of 100 µl; the reaction mixture contained 50 mM MES, 0.5 mM $MgCl_2$, 1 mM EGTA, 250 µM [γ-^{32}P]ATP , 1000 dpm/pmol and 500 µg of microtubular protein. After 15 min of incubation at 30°C, 18 µl were taken and added to 40 µl of sample buffer (125 mM Tris–HCl, pH 6.8, containing 20% glycerol, 10% ß–mercaptoethanol and 4.6% SDS) . Samples were heated for 5 min in a boiling water bath and subjected to SDS–polyacrylamide electrophoresis, using 10% gel. The radioactive proteins were visualized by fluorography. C = microtubules from control rats; A = microtubules from hyperammonemic rats.

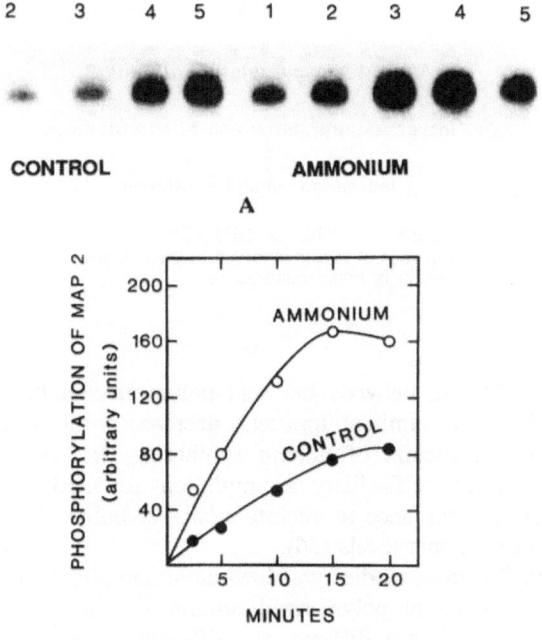

Figure 2. Kinetics of phosphorylation of MAP-2 in isolated microtubules from control and hyperammonemic rats. A. Rats were fed control or ammonium diet for 15 days and microtubules were isolated from brain. Protein phosphorylation was assayed as in Fig. 1. 15 μl of the incubation mixture were taken at 2.5 (1), 5 (2), 10 (3), 15 (4) and 20 (5) min and subjected to electrophoresis and fluorography. The band corresponding to MAP-2 is shown. B The experiment shown in A was repeated four times with two different preparations of microtubules. The intensities of the spots corresponding to MAP-2 in the fluorography were quantified using a laser densitometer. The mean values are given.

tubulin in brain (17, 18). Tubulin is the basic constituent of microtubules, one of the main components of the cytoskeleton. Microtubules play important roles in a lot of cellular functions, including control of cell architecture, shape, motility, mitosis and intracellular organelle movement (19, 20). Tubulin is specially abundant in brain because axons contain large quantities of microtubules which serve as a guide for axonal transport of organelles including mitochondria and neurosecretory vesicles. Thus, microtubules play an important role in neurotransmission and alterations in their amount or functionallity could lead to neurologic disfuntions such as those found in hepatic encephalopathy. It was therefore considered of interest to caracterize the process involved in the induction of tubulin in hyperammonemia. The effect is completely reversible following withdrawal of the ammonium diet. Tubulin is selectively induced in certain areas of the brain (21, 22) with no effect in non neural tissues (18). Moreover, tubulin content increases in neurons but not in astrocytes.

Tubulin is the main component of the microtubules but is also present within the cells in a free form. There is a dynamic equilibrium of tubulin between the free and polymerized forms. In mammalian cultured cells, it has been shown that agents which increase free tubulin levels inhibit tubulin synthesis, while those that decrease free tubulin increase tubulin synthesis (23). The synthesis of tubulin is thus mainly controlled by the extent of its polymerization. The effect of hyperammonemia on tubulin polymerization was therefore assessed. The content of free tubulin was not significantly affected while that of polymerized tubulin increased remarkably (24); moreover, changes in tubulin polymerization precede those in tubulin synthesis (25). These results indicate that the primary effect produced by ammonia

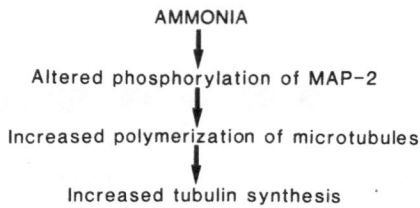

AMMONIA

Altered phosphorylation of MAP-2

Increased polymerization of microtubules

Increased tubulin synthesis

Figure 3. SEQUENCE OF EVENTS. Proposed sequence of events involved in the induction of tubulin in hyperammonemia.

is a shift of the equilibrium between free and polymerized tubulin, leading to increased polymerization, with a concomitant, transient, decrease in the content of free tubulin and, as a consequence, an induction of tubulin synthesis. The increased polymerization of microtubules could explain the fibrillary accumulations reported in 1971 by Cavanagh et al, which suggested that a disturbance to microtubular metabolism in the brain could occur in association with high ammonia levels (26).

The next question to be studied was how ammonia affects polymerization of tubulin. Microtubules are composed of polymerized tubulin and microtubule–associated proteins (MAPs). The pattern of MAPs is different for different tissues and cell types. In neurons, polymerization of microtubules is mainly controlled by two sets of MAPs, i. e. MAP–2 and Tau proteins. These MAPs stimulate polymerization of microtubules (27–29) and the stimulatory capacity varies with the state of phosphorylation of these MAPs (30–33). We therefore studied if hyperammonemia affects the amount or the phosphorylation of these MAPs in brain.

Hyperammonemia did not affect the amount of MAPs in the whole brain (24). The effect of hyperammonemia on the phosphorylation of microtubule–associated proteins was studied in vitro using microtubules isolated from control or hyperammonemic rats. As shown in Fig. 1, hyperammonemia clearly affects the pattern of proteins phosphorylated. Using the same amount of total microtubular protein, the phosphorylation of some proteins remains unchanged in hyperammonemic rats while for other proteins, indicated by arrows and including MAP–2, in vitro phosphorylation was selectively and remarkably increased in microtubules from hyperammonemic rats. The kinetics of phosphorylation of MAP–2 are shown in Fig. 2. These results suggest that hyperammonemia affects phosphorylation of some MAPs, including MAP–2. The altered phosphorylation of MAP–2 would led to increased polymerization of microtubules which in turn would induce tubulin synthesis.

MAP–2 is localized specifically in neurons and plays a key role in the control of tubulin polymerization in neurons. Therefore, the sequence of events proposed above and in Fig. 3 could explain the fact that the increase in tubulin induced by hyperammonemia is selective for neurons.

3. Acute Ammonium Injection Induces a Proteolysis of MAP-2

It was then considered of interest to study the mechanism involved in the alteration of MAP–2 phosphorylation in hyperammonemia and also that the mechanism could be better studied using acute ammonia intoxication.

Rats were injected intraperitoneally with a non lethal dose (6 mmol/Kg) of ammonium acetate or with saline (controls). As shown in Fig. 4, microtubules isolated from brain two hours after ammonium injection show a reduced content of MAP–2. This decrease could be due to a reduction of the total content of MAP–2 in brain by proteolysis or to decreased

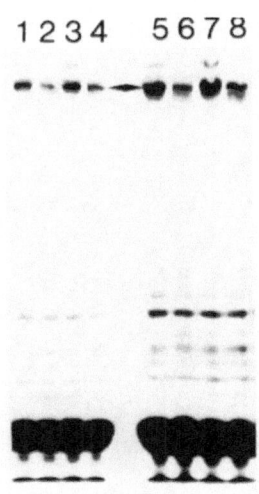

Figure 4. Ammonium injection reduces the content of MAP–2 in microtubules. Rats were injected i.p. with 6 mmol/Kg of ammonium acetate or saline (controls). After 2 hours rats were killed and brain microtubules were isolated and subjected to SDS–polyacrylamide gel electrophoresis. Microtubules isolated from two different rats per group are shown. 50 (lanes 1–4) or 100 (lanes 5–8) µg of total microtubule protein were applied to the gel. Arrows indicate the position of MAP–2. From 34.

binding of MAP–2 to tubulin with no effect on the total amount of MAP–2. We determined by immunoblotting the content of MAP–2 in whole homogenates from brain; as shown in Fig. 5, the total amount of MAP–2 in brain is remarkably reduced in rats injected with ammonium acetate, indicating the induction of a selective proteolysis of MAP–2. The time–course of MAP–2 degradation is shown in Fig. 6.

4. Ammonia–Induced Proteolysis of MAP–2 is Mediated by Activation of the NMDA Receptor

It has been reported that activation of the NMDA receptor leads to dephosphorylation of MAP–2 (35). Also, glutamatergic neurotransmission is altered in hyperammonemia (36–40), with selective loss of NMDA–sensitive glutamate binding sites in rat brain following portacaval anastomosis (40). Feeding rats the ammonium containing diet to produce chronic hyperammonemia also induces alterations in glutamatergic neurotransmission. Moreover, cerebral ischemia, which is associated with increased release of glutamate, leads to the rapid disappearance of MAP–2 (41). We suspected therefore that the effects of hyperammonemia on MAP–2 could be mediated by activation of glutamate receptors.

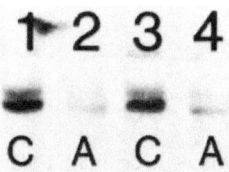

Figure 5. Ammonium injection induces proteolysis of MAP–2 in brain. Rats were injected with ammonium acetate as in Fig. 4. After two hours, rats were killed and the brains were homogenized. 40 µg of the homogenate proteins were subjected to SDS–polyacrylamide electrophoresis and immunoblotting using an antibody to MAP–2. Samples corresponding to two different rats per group are shown. C = brain homogenates from control rats; A = brain homogenates from rats injected with ammonium acetate. From 34 .

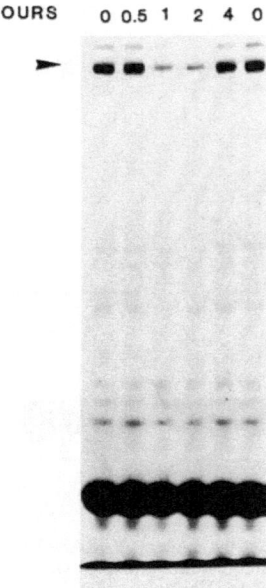

HOURS 0 0.5 1 2 4 0

Figure 6. Time–course of the degradation of MAP–2 in rats injected with ammonium acetate. Rats were injected i.p. with 6 mmol/Kg of ammonium acetate and killed after 0, 0.5, 1, 2 or 4 hours. Brain microtubules were isolated and subjected to electrophoresis as in Fig. 4. 40 μg of protein was applied to each lane. Arrow indicates the position of MAP–2. From 34 .

To assess whether the ammonia–induced proteolysis of MAP–2 is mediated by activation of the NMDA receptor, we injected rats with MK–801, a specific antagonist of this receptor, before injecting ammonia. As shown in Fig. 7, MK–801 completely prevents the proteolyisis of MAP–2. This suggests that ammonia injection leads to an activation of the NMDA receptor which in turn mediates the activation of a protease which selectively degrades MAP–2. L–carnitine, which prevents ammonia toxicity (42), also prevents ammonia–induced proteolysis of MAP–2 (Fig. 7).

Activation of the NMDA receptor induces a rise in free intracellular Ca^{+2}. We therefore assessed whether brain homogenates contain a Ca^{+2}–dependent protease which could

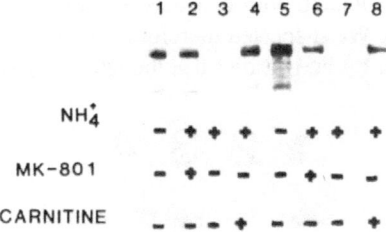

| | 1 | 2 | 3 | 4 | 5 | 6 | 7 | 8 |

NH_4^+ — + + + — + + +

MK–801 — + — — — + — —

CARNITINE — — + — — — — +

Figure 7. Protective effect of MK–801 and of L–carnitine against ammonium–induced proteolysis of MAP–2. Groups of two rats were injected with saline or with 7 mmol/Kg of ammonium acetate. Two rats were injected with 2 mg/Kg MK–801 15 min before ammonium injection. Another group of two rat was injected with 16 mmol/Kg L–carnitine 1 hour before ammonium injection. Rats were killed 90 min after ammonium injection. Brains were removed immediately and homogenized. Aliquots of the homogenate (30 μg of protein) were subjected to electrophoresis and MAP–2 was visualized by immunoblotting. The figure shows the results for two different rats per group. Lanes 1 and 5: control rats; lanes 3 and 7: rats injected only with ammonium acetate; lanes 2 and 6: rats injected with MK–801 and ammonium acetate; lanes 4 and 8: rats injected with L–carnitine and ammonium acetate. From 34 .

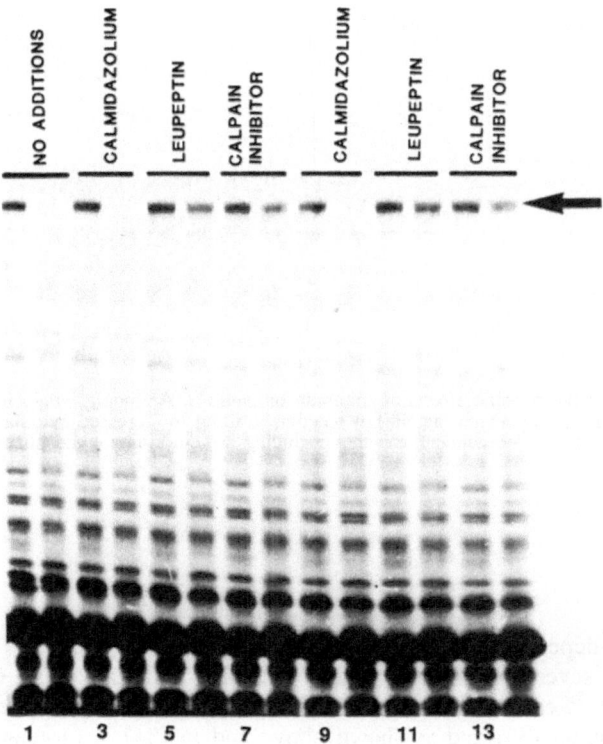

Figure 8. Ca^{+2}–dependent proteolysis of MAP–2 in brain supernatants is inhibited by inhibitors of calpain I. Rats were killed and brains were homogenized in 9 ml/g of 250 mM sucrose, 5 mM $MgCl_2$, 50 mM Tris–HCl, pH 7.4. The homogenate was centrifuged at 100,000 g for 1 hour at 4°C. The supernatant was used as the source of protease for the assay. The final assay contained the following in a volume of 60 µl: 50 µg of supernatant protein, isolated microtubules (50 µg of protein), 50 mM Tris–HCl, 3 mM $MgCl_2$, 100 mM sucrose, pH 7.4 and the indicated amounts of inhibitors. $CaCl_2$, 2 mM final concentration was added to the samples run in the even lanes; the odd lanes are the corresponding controls incubated without the addition of Ca^{+2}. Calmidazolium was added to the samples run in lanes 3 and 4 (20 µM calmidazolium) and lanes 9 and 10 (40 µM). Leupeptin was added to samples in lanes 5 and 6 (20 µg/ml) and lanes 11 and 12 (40 µg/ml). A specific inhibitor of calpain I (calpain inhibitor I, Boehringer Mannheim) was added to samples in lanes 7 and 8 (17 µg/ml) and lanes 13 and 14 (34 µg/ml). Samples were incubated for 2 hours at 37°C, and then subjected to electrophoresis. Proteins were stained with Coomassie Blue. From 34.

degrade MAP–2. As shown in Fig. 8, MAP–2 is selectively degraded by a cytosolic, Ca^{+2}–dependent, protease which can be inhibited by leupeptin and by a specific inhibitor of calpain I. It can be seen that, under the conditions used, no other protein is significantly degraded. These results suggest that MAP–2 is selectively degraded by a Ca^{+2}–dependent protease and that, presumably, this protease would be calpain, because it is inhibited by a calpain inhibitor.

5. Acute Ammonia Toxicity is Mediated by Activation of the NMDA Receptor

The above results show that ammonium injection induces a proteolysis of MAP–2 that is mediated by activation of the NMDA type of glutamate receptors. This suggests that ammonium induces a release of glutamate which in turn activates the NMDA receptor. This would be in agreement with previous reports showing that i.p. injection of ammonium acetate induces a release of glutamate in the cortex of rat in vivo (36). Following activation of the NMDA receptor, there should be an increase in $[Ca^{+2}]_i$, which, in turn, should lead to

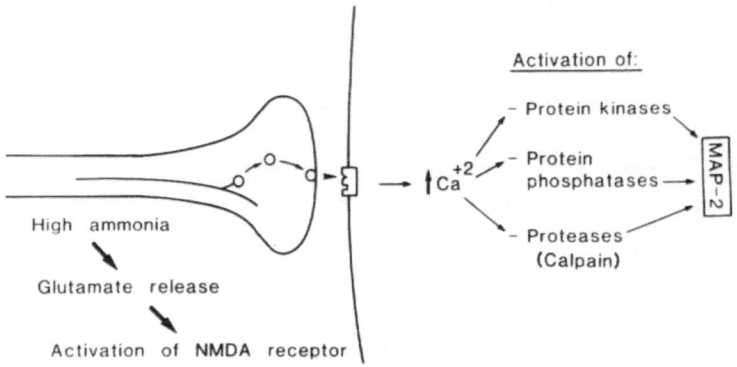

Figure 9. Scheme of the possible effects of ammonia on neurons. Ammonia would increase the release of glutamate, which would activate the NMDA receptor, leading to increased cytoplasmic calcium; this increase would activate Ca^{+2}–dependent enzymes including protein kinases, phosphatases and proteases.

activation of Ca^{+2}–dependent protein kinases, phosphatases and proteases (Fig. 9). MAP–2 is a substrate for several of these enzymes including protein kinase C, protein kinase dependent on Ca^{+2} and calmodulin (43, 44), a phosphatase which is activated upon activation of the NMDA receptor (35) and as shown above and in (34), a protease which could be calpain I.

Our results suggest that in chronic, mild hyperammonemia, there is a slight alteration in the function of the NMDA receptor, leading to changes in the activities of protein kinases and/or phosphatases (Fig. 9) which would result in an altered extent of MAP–2 phosphorylation. In contrast, following acute intoxication with high doses of ammonia, there is a stronger activation of the NMDA receptor, leading to activation of a protease which selectively degrades MAP–2.

It is well known that glutamate is neurotoxic; persistent stimulation of glutamate receptors causes neuronal death. Glutamate induces a receptor–mediated rise in the cytosolic free Ca^{+2} concentration (45). A large body of evidence has associated delayed glutamate-induced neuronal death with sustained increases in $[Ca^{+2}]_i$ (46, 47). It has also been clearly stablished that antagonists of the NMDA receptor prevent neuronal death induced by glutamate (48), suggesting that the rise of $[Ca^{+2}]_i$ induced by activation of the NMDA receptor is a key step in the process of neuronal death.

As shown above and summarized in Fig. 9, acute ammonium injection induces an activation of the NMDA receptor. We therefore considered of interest to assess if acute ammonia toxicity could be mediated by activation of this receptor, as is the case for glutamate neurotoxicity. To assess this posibility we tested if a highly specific antagonist of the NMDA receptor (MK–801) is able to prevent ammonia toxicity. Groups of mice were injected i.p. with 12 mmol/Kg of ammonium acetate. As shown in Table 1, 73% of the mice died. When mice were injected with MK–801 (2 mg/Kg) 15 min before ammonium injection, only 5% of the animals died. A similar experiment was carried out with rats; groups of rats were injected i.p. with 7 mmol/Kg of ammonium acetate. 70% of the rats died. However, when the rats were previously injected with MK–801, only 15% died (49). This results clearly show that MK–801 protects the animals against acute ammonia toxicity, indicating that it is mediated by activation of the NMDA receptor. This fits well with the scheme shown in Fig. 9; an excessive activation of the NMDA receptor would lead to neuronal death and finally to the death of the animal.

Table 1. Protective effect of MK–801 against ammonia toxicity.

	MK–801	Animals injected	Survivors	Survival (%)
Mice	NO	41	11	27
	YES	20	19	95
Rats	NO	31	10	32
	YES	31	26	84

Mice and rats were injected I.P. with 12 and 7 mmol/Kg, respectively, of ammonium acetate. The experimental groups were injected i.p. with 2 mg/Kg of MK–801 15 min before injecting ammonium acetate. From 49.

6. Concluding Remarks

In summary, the results discussed above indicate that chronic, mild hyperammonemia, affects microtubule polymerization in brain likely by altering the extent of phosphorylation of the microtubule–associated protein MAP–2. Acute injection of high doses of ammonia induces a selective proteolysis of MAP–2 in rat brain. This proteolysis seems to be mediated by activation of the NMDA type of glutamate receptors which results in increased cytoplasmic Ca^{+2} and activation of a Ca^{+2}-dependent protease (presumably calpain) which selectively degrades MAP–2. The results shown in Table 1 indicate that acute ammonia toxicity is mediated by activation of the NMDA receptor. This points out the molecular mechanism involved in ammonia toxicity. If this hypothesis is further confirmed by additional experiments, it could be of great interest from the basic scientific point of view. Moreover the clinical implications of this hypothesis are also of evident interest.

REFERENCES

1. Bessman, S. P., and Bessman, A. N., 1955, The cerebral and peripheral uptake of ammonia in liver disease with an hypothesis for the mechanism of hepatic coma, J. Clin. Invest. **34**:622–628.
2. Fischer, J. E., and Baldessarini, R. J., 1971, False neurotransmitters and hepatic failure, Lancet. **II**:75–79.
3. James, J. H., Ziparo, V., Jeppson, B., and Fischer, J. E., 1979, Hyperammonemia, plasma amino acid imbalance, and blood–brain amino acid transport: a unified theory of portal–systemic encephalopathy, Lancet. **II**:772–775.
4. Zieve, L., Doizaki, W. M., and Zieve, F. J., 1974, Synergism between mercaptans and ammonia in the production of coma: a possible role for mercaptans in the pathogenesis of hepatic coma, J. Lab. Clin. Med. **83**:16–28.
5. Schafer, D. F., and Jones, E. A., 1982, Hepatic encephalopathy and the þ–aminobutyric acid neurotransmitter system, The Lancet. **II**:18–19.
6. Bircher, J., Müller, J., Guggenheim, P., and Haemmerli, U. P., 1966, Treatment of portal–systemic encephalopathy with lactulose, Lancet. **I**:890–893.
7. Hahn, M., Massen, O., Nencki, M., and Pawlow, J., 1893, Die Eck'sche Fistel zwischen der unteren Hohlvene und der Pfortader und ihre Folgen für den Organismus. Archiv. f. Experiment. Pathol. u, Pharmakol. **32**:161–210.
8. Eck, N. V., 1877, Kvoprosu o perevyazkie vorotnois veni. Predvaritelnoye soobshtshjenye, Voen. Med. J. St. Petesburg. 130, 2:1–2. English translation: Child, C. G. 1953, Concerning ligation of the vena porta, Surgery, Gynecol. and Obstetr. **96**:375–376.
9. Lee, S. H., and Fischer, B., 1961, Portacaval shunt in the rat, Surgery. **50**:668–672.
10. Kyu, M. H., and Cavanagh, J. B., 1970, Some effects of porto–caval anastomosis in the male rat, Br. J. Exp. Path. **51**:217–227.

11. Blitzer, B. L., Waggoner, J. G., Jones, E. A., Gralnick, H. R., Towne, D., Butler, J., Weise, V., Kopin, I. J., Walters, I., Teychenne, P. F., Goodman, D. G. and Berk, P. D. 1978, A model of fulminant hepatic failure in the rabbit, Gastroenterology. **74:**664–671.

12. Kelly, J. H., Koussayer, T., He, D–E., Chong, M. G., Shang, T. A., Whisennand, H. H., and Sussman, N. L., 1992, An improved model of acetaminophen–induced fulminant hepatic failure in dogs, Hepatology. **15:**329–335.

13. Fick, T. E., Schalm, S. W., and de Vlieger, M., 1989, Continuous intravenous ammonia infusion as a model for the study of hepatic encephalopathy in rabbits, J. Surg. Res. **46:**221–225.

14. Clifford, A. J., Prior, R. L., and Visek, W. J., 1969, Depletion of reduced pyridine nucleotides in liver and blood with ammonia, Am. J. Physiol. **217:**1269–1272.

15. Felipo, V., Miñana, M. D., and Grisolía, S., 1988, Long–term ingestion of ammonium increases acetylglutamate and urea levels without affecting the amount of carbamoyl–phosphate synthase, Eur. J. Biochem. **176:**567–571.

16. Azorín, I., Miñana, M. D., Felipo, V., and Grisolía, S., 1989, A simple animal model of hyperammonemia, Hepatology. **10:**311–314.

17. Felipo, V., Miñana, M. D., Azorín, I., and Grisolía, S., 1988, Induction of rat brain tubulin following ammonium ingestion. J. Neurochem. **51:**1041–1045.

18. Miñana, M. D., Felipo, V., and Grisolía, S., 1990, Hyperammonemia induces brain tubulin, In: Cirrhosis, Hepatic Encephalopathy and Ammonium Toxicity Grisolía, S., Felipo, V., and Miñana, M. D., eds, Adv. Exp. Med. Biol. **272:**65–80. Plenum Press, New York.

19. Artvinli, S., 1987, Cytoskeleton, microtubules, tubulin and colchicine: A review. Cytologia. **52:**189–198.

20. Weiss, D. G., Seith–Tutter, D., Langford, G. M., and Allen, R. D., 1987, The native microtubule as the engine for bidirectional organelle movement, In Axonal Transport, Alan R. Liss, Inc. 91–111.

21. Miñana, M. D., Felipo, V., Wallace, R., and Grisolía, S., 1988, High ammonia levels in brain induce tubulin in cerebrum but not in cerebellum, J. Neurochem. **51:**1839–1842.

22. Miñana, M. D., Felipo, V., Quel, A., Pallardó, F., and Grisolía, S., 1989, Selective regional distribution of tubulin induced in cerebrum by hyperammonemia, Neurochem. Res. **12:**1241–1243.

23. Ben–Ze'ev, A., Farmer, S. R., and Penman, S., 1978, Mechanisms of regulating tubulin synthesis in cultured mammalian cells, Cell. **17:**319–325.

24. Miñana. M. D., Felipo, V., and Grisolía, S., 1989, Assembly and disassembly of brain tubulin is affected by high ammonia levels, Neurochem. Res. **14:**235–238.

25. Felipo, V., Miñana, M. D., and Grisolía, S., 1990, Hyperammonemia induces polymerization of brain tubulin, Neurochem. Res. 15:945–948.

26. Cavanagh, J. B., Blakemore, W. F., and Kyu, M. H., 1971, Fibrillary accumulations in oligodendroglial processes of rats subjected to portacaval anstomosis, J. Neurol. Sci. **14:**143–152.

27. Murphy, D. B., and Borisy, G. C., 1975, Association of high molecular weight proteins with microtubules and their role in microtubule assembly in vitro, Proc. Natl. Acad. Sci. USA **72:**2696–2700.

28. Weingarten, M. D., Lockwood, A. H., Hwo, S, Y., and Kirschner, M. W., 1975, A protein factor essential for microtubule assembly, Proc. Natl. Acad. Sci. USA. **72:**1858–1862.

29. Kim, H., Binder, L. I., and Rosenbaum, J. I., 1979, The periodic association of MAP$_2$ with brain icrotubules in vitro, J. Cell Biol. **80:**266–276.

30. Jameson, L., Frey, T., Zeeberg, B., Dalldorf, F., and Caplow, M., 1980, Inhibition of microtubule assembly by phosphorylation of microtubule–associated proteins, Biochemistry. **19:**2472–2479.

31. Jameson, L., and Caplow, M., 1981, Modification of microtubule steady–state dynamics by phosphorylation of the microtubule–associated proteins, Proc. Natl. Acad. Sci. USA. **78:**3413–3417.

32. Yamamoto, H., Saitoh, Y., Fukunaga, K., Nishimura, H., and Miyamoto, E., 1988, Dephosphorylation of microtubule proteins by brain protein phosphatases 1 and 2A, and itseffect on microtubule assembly, J. Neurochem. **50:**1614–1623.

33. Bruggs, B., and Matus, A., 1991, Phosphorylation determines the binding of microtubule–

associated protein 2 MAP2 to microtubules in living cells, J. Cell. Biol. **114:**735–743.

34. Felipo, V., Grau, E., Miñana, M. D., and Grisolía, S., 1993, Ammonium injection induces an N–methyl–D– aspartate receptor–mediated proteolysis of the microtubule–associated protein MAP-2. J. Neurochem. in press.

35. Halpain, S., and Greengard, P., 1990, Activation of. NMDA receptors induces rapid dephosphorylation of the cytoskeletal protein MAP2, Neuron. **5:**237–246.

36 Moroni, F., Lombardi, G., Moneti, G., and Cortesini, C., 1983, The release and neosynthesis of glutamic acid are increased in experimental models of hepatic encephalopathy, J. Neurochem. **40:**850–854.

37. Ferenci, P., Pappas, S. C., Munson, P. J., and Jones, E. A., 1984, Changes in glutamate receptors on synaptic membranes associated with hepatic encephalopathy or hyperammonemia in the rabbit, Hepatology. **4:**25–29. 38. Rao, V. L. R., and Murthy, C. R. K., 1991, Hyperammonemic alterations in the uptake
and//release of glutamate and aspartate by rat cerebellar preparations, Neurosci. Lett. **130:**49–52.

39. Rao, V. L. R., Murthy, C. R. K., and Butterworth, R. F., 1992, Glutamatergic synaptic dysfunction in hyperammonemic syndromes, Metabol. Brain Dis. **7:**1–20.

40. Peterson, C., Giguere, J. J., Cotman, C. W., and Butterworth, R. F., 1990, Selective loss of N–methyl–D–aspartate–sensitive L–[^{3}H]glutamate binding sites in rat brain following portacaval anastomosis, J. Neurochem. **55:**386–390.

41. Kitagawa, K., Matsumoto, M., Niinobe, M., Mikoshiba, K., Hata, R., Ueda, H., Handa, N., Fukunaga, R., Isaka, I., Kimura, K., and Kamada, T., 1989, Microtubule–associated protein 2 as a sensitive marker for cerebral ischemic damage. Immunohystochemical investigation of dendritic damage, Neuroscience. **31:**401–411.

42. O'Connor, J. E., Costell, M., and Grisolía, S., 1984, Protective effect of L–carnitine on hyperammonemia. FEBS Lett. **166:**331–334.

43. Yamauchi, T., and Fugisawa, H., 1982, Phosphorylation of microtubule–associated protein 2 by calmodulin–dependent protein kinase kinase II which occurs only in brain tissues, Biochem. Biophys. Res. Commun. **109:**975–981.

44. Walaas, S. I., and Nairn, A. C., 1989, Multisite phophorylation of microtubule–associated protein 2 MAP-2 in rat brain: peptide mapping distinguishes between cyclic AMP–, calcium/calmodulin–, and calcium/phospholipid–regulated phosphorylation mechanisms, J. Mol. Neurosci. **1:**117–127.

45. MacDermot, A. B., Mayer, M. L., Westbrook, G. L., Smith, S. J., and Barker, J. L., 1986, NMDA receptor activation increases cytoplasmic calcium concentration in cultured spinal cord neurones, Nature, **321:**519–522.

46. Choi, D. W., 1985, Glutamate neurotoxicity in cortical cell culture is calcium– dependent, Neurosci. Lett. **58:**293–297.

47. Manev, H., Favaron, M., Guidotti, A., and Costa, E., 1989, Delayed increase of Ca^{+2} influx elicited by glutamate: role in neuronal death, Mol. Pharmacol. **36:**106–112.

48. Olney, J., Price, M., Salles, K. S., Labruyere, J., and Friedrich, G., 1987, MK–801 powerfully protects against N–methyl–D–aspartate neurotoxicity, Eur. J. Pharmacol. **141:**357–361.

49. Marcaida, G., Felipo, V., Hermenegildo, C., Miñana, M. D., and Grisolía, 1992, Acute ammonia toxicity is mediated by the NMDA type of glutamate receptors, FEBS Lett. **296:**67–68.

Modulation of the Exocytotic Release of Neurotransmitter Glutamate by Protein Kinase C

J. Sánchez–Prieto, I. Herrero and M.T. Miras–Portugal

1. Introduction

Most excitatory synapses in the brain use the amino acid glutamate as their transmitter. Glutamate neurotransmission, being so wide spread, is involved in many functions in the Central Nervous System. Thus, there is evidence that an increase in the effectiveness of glutamatergic synapses mediate some forms of synaptic plasticity which participate in processes of memory and learning. On the other hand, glutamate is involved in the brain damage associated with hypoxia, hypoglycaemia, ischaemia, epilepsy and some neurodegenerative diseases.

The stablishment of the compartmentation of glutamate within the nerve terminal has provided a better understanding of the mechanism by which excitatory amino acid are released into the synaptic cletf. In addition, the discovery of different pathways for the transduction of extracellular signals has also help to understand the way in which synaptic transmission can be modulated through the control of the amount of neurotransmitter that is released by the nerve ending. A cellular signalling pathway involves the generation in a receptor–dependent manner of the intracellular messengers inositol trisphosphate, IP_3, and diacylglycerol. IP_3 releases Ca^{2+} from intracellular stores whereas diacylglycerol activates a protein kinase dependent of Ca^{2+} and phospholipids, PKC. In nerve cells, PKC is involved in the control of neurotransmitter release through the modulation of the activity of ionic channels. In this paper we will review some aspects of the role of PKC in the regulation of glutamate exocytosis in nerve terminals and the receptors involved in the activation of this cellular mechanism for signal transduction.

2. Glutamate Release In Nerve Terminals

Isolated nerve terminals (synaptosomes) are obtained by the homogenization of brain regions such as the cerebral cortex and the hippocampus. On isolation, synaptosomes retain the biochemical machinery for the uptake, store, and release of neurotransmitters, thus providing the simplest model in which the mechanism for transmitter release can be studied (1).

Departamento de Bioquímica, Facultad de Veterinaria, Universidad Complutense, Madrid 28040, Spain.

Cirrhosis, Hyperammonemia, and Hepatic Encephalopathy,
Edited by S. Grisolia and V. Felipo, Plenum Press, New York, 1994

Within the nerve terminal glutamate is accumulated across the plasma membrane by a Na^+–dependent acidic amino acid carrier also present in glial plasma membranes, that translocate not only L–glutamate but also L–aspartate and D–aspartate, (2). Under resting conditions, the glutamate carrier maintains in the extracellular medium a concentration of glutamate of 1 μM, to prevent the neurotoxic actions of the amino acid. Cytoplasmic glutamate is accumulated in small synaptic vesicles by means of a transporter, highly specific for glutamate , and dependent on the electrochemical gradient maintained by a vesicular ATP ase (3).

The mechanism by which glutamate is released can be studied with a fluorometric technique by incubating nerve terminals in the presence of NADP and glutamate dehydrogenase (4,5). The glutamate released from the nerve ending is oxidized to 2–oxoglutarate by the enzyme and the NADP reduced to NADPH. Depolarization of synaptosomes with KCl or with the K^+–channel blocker, 4–aminopyrydine, 4AP, has shown a Ca^{2+}–dependent and a Ca^{2+}–independent components for glutamate release. (6). However, the Ca^{2+}–independent component was reduced during the transient depolarizations induced by 4AP.

The Ca^{2+}–dependent release of glutamate originates from a non–cytoplasmic vesicular compartment, is highly dependent on metabolic energy, (7,8) and is blocked by the incubation with botulinum toxin A (9). The exocytotic release of glutamate is triggered by the localized increase in the intraterminal concentration of Ca^{2+} (10) that follows depolarization and the opening of voltage sensitive Ca^{2+}–channels. KCl–induced exocytosis of glutamate is byphasic (8) with a rapid phase that reflects the release of the synaptic vesicles already docked at the active zone, followed by a slow phase consistent with the translocation of more vesicles to the release site.

In contrast to the Ca^{2+}–dependent, the Ca^{2+}–independent component of glutamate release is mediated by the reversal of the electrogenic acidic amino acid transporter. On depolarization, the carrier releases cytoplasmic glutamate in amounts correlating with the extent and the duration of the collapse of the Na^+–electrochemical gradient. The Ca^{2+}–independent release of glutamate does not require energy. It is in fact, stimulated by the decrease in the intraterminal ATP/ADP ratio that follows the lack of oxygen in the incubation medium or the presence of metabolic inhibitors (11–13). The amount of glutamate released by reversal of the glutamate carrier during "in vitro" depolarizations is likely to be reduced during an "in vivo" action potential. However, in pathological conditions such as hypoxia and ischaemia, where a chronic decrease in the Na^+–electrochemical gradient occurs as a consequence of the fall in ATP and the reduced activity of Na^+-K^+-ATPase, the release of cytoplasmic glutamate through the Ca^{2+}–independent pathway can lead to the accumulation of excitotoxic concentrations of glutamate in the extracellular medium. Glutamate neurotoxicity is mediated, at least in part, by the massive accumulation of Ca^{2+} in the cell that follows a excessive stimulation of the glutamatergic, N–methyl–D–aspartate, NMDA, receptor (14).

2.1 PKC In Nerve Terminals

Protein kinase C, PKC, is a phospholipid and Ca^{2+}–dependent protein kinase which is particularly abundant in the cerebral tissues (15). In the brain, this enzyme plays an important role in several neuronal functions such as the modulation of neurotransmitter release and ionic channel activities, long term potentiation and neuronal development and regeneration(16–18) PKC is known to be a large family of proteins integrated by different subspecies. Initially, four cDNA clones encoding for the α, $β_I$, $β_{II}$ and γ subspecies were found. Subsequently, a second group of cDNA clones encoding for the δ,ε,ζ, and η subspecies were identified.

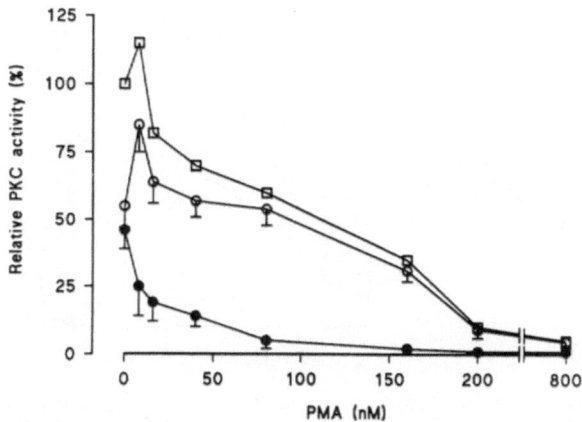

Figure 1. Translocation of protein kinase C activity in cerebrocortical synaptosomes exposed to phorbol myristoil acetate (PMA). Synaptosomes were incubated for 10 min in the presence of increasing concentrations of PMA. After homogenization and centrifugation the protein kinase C was purified as described (26) and the activity, estimated as the 32P incorporated into histone H_1, assayed in the soluble, cytosolic fraction (●) and the particulate membrane fraction (○); (□), total activity. The total activity in the absence of PMA was taken as 100%. Results are means ± SEM of three experiments. [From Díaz–Guerra et al. (26)]

These enzyme subtypes differ in many respects such as mode of activation, sensitivity to Ca^{2+}, catalitic activity toward endogenous substrates, tissue expression and rate of down regulation. (19).

PKC is activated by the diacylglycerol generated by a receptor mediated increase in the metabolism of phosphatidylinositol. Activation of PKC involves its translocation from a cytoplasmic inactive form, prevalent at resting conditions, to a membrane bound form. The subcellular redistribution of PKC can be induced also by an increase in the intracellular concentration of Ca^{2+}. Thus, PKC binds to phosphatidylserine and becomes membrane bound but not active (20). The diacylglycerol generated by receptor activation binds to phosphatidylserine–Ca^{2+}–PKC complex to produce the active form (21). PKC is also translocated to the membrane and activated by phorbol esters at resting Ca^{2+} concentrations

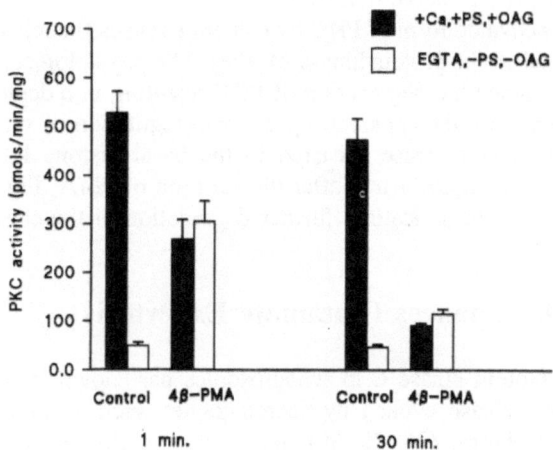

Figure 2. Transient appearence of a protein kinase activity independent of Ca^{2+} and phospholipids during the down regulation of synaptosomal PKC with phorbol esters. Synaptosomes were incubated with 1 µM PMA and the PKC activity extracted with nonidet 0.5 %. The suspension was centrifuged at 140,000 x g for 30 min and the PKC activity was assayed in the supernatants without any further purification. Results are means of three experiments ± SEM.

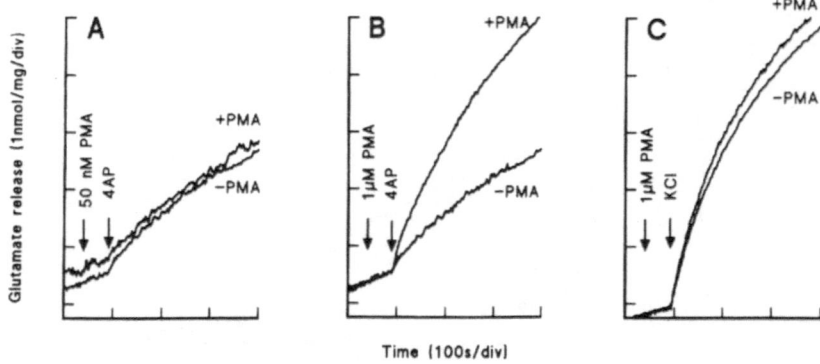

Figure 3. High concentrations of PMA enhance the 4-aminopyridine, 4AP-evoked but not the KCl-evoked release of glutamate. Glutamate release was determined by the fluorometric technique described in (5) in the presence of 1.33 mM $CaCl_2$. Glutamate release was evoked by depolarization of synaptosomes with 50 μM 4AP (A) and (B) or with 30 mM KCl (C).

(22). PMA-induced translocation involves the insertion of the enzyme into the lipid bilayer resulting initially in activation but followed by down regulation (23).

Inmunocytochemical studies of cerebral cortex have shown that PKC is present in the presynaptic nerve terminals (24). Synaptosomes prepared from this cerebral region contain the α,β and γ subspecies while synaptosomes from the hippocampus or from the cerebellar cortex contain only the α and β subspecies (25). Studies on the subcellular distribution of PKC have shown that in unstimulated synaptosomes a 50% of PKC activity is found associated with the membrane fraction (26). Depolarization of synaptosomes with KCl does not cause any redistribution of PKC. However, an extensive translocation of the enzyme to the membrane fraction is caused either by the synthetic diacylglycerol oleoylacetylglycerol (OAG) or the phorbol ester phorbol myristoyl acetate (PMA), Figure 1. Low concentrations of PMA , extensively translocate PKC from the soluble to the particulate fraction as judged by the reciprocal changes in the soluble and particulate activities. Half maximal decrease in cytoplasmic activity was obtained with 10 nM PMA with an apparently low activity being detected at 800 nM PMA. These results are consistent with the rapid down-regulation of the α and β subspecies of PKC in nerve terminals (27).

Proteolysis of the activated form of PKC by calpain I causes the release of a catalytically fully active fragment, called the protein kinase M, (28). The physiological significance of this proteolysis may be to initiate the degradation of PKC resulting in a depletion of the enzyme from the cell. Consistent with the apparent rapid down-regulation of synaptosomal PKC by phorbol esters, a significant increase occurred in the basal histone kinase activity in the absence of Ca^{2+} and phospholipids 1 min after the addition of PMA, Figure 2. This increase in PKM activity was transient indicating further degradation of the enzyme by proteases.

3. PKC Activation Enhances Glutamate Exocytosis

The activation of protein kinase C in synaptosomes has shown a marginal stimulation of the neurotransmitter release evoked by secretagogues such as high extracellular KCl, veratridine or Ca^{2+} ionophores. (26, 29, 30), in contrast to the two to three fold increase observed in electrically stimulated brain slices. (31,32). However, the finding that 4-aminopyridine induces spontaneous and tetrodotoxin-sensitive action potentials (33), has provided a more physiologycal way to stimulate the nerve terminals in order to induce transmitter release. Thus, in 4AP-stimulated synaptosomes the activation of PKC with phorbol esters greatly enhances the release of glutamate without significant effects either in

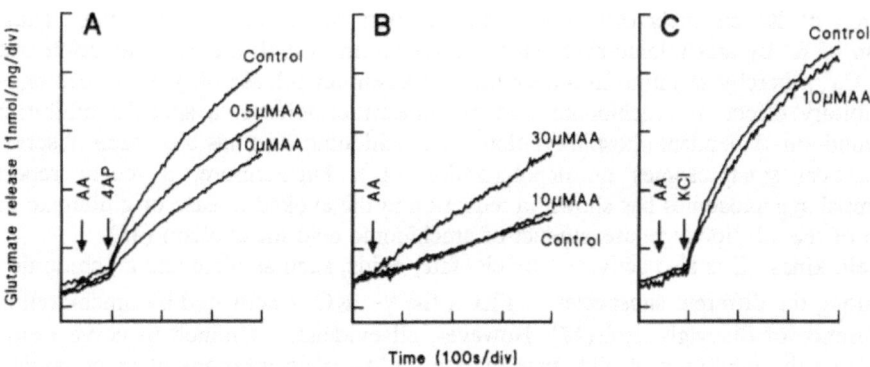

Figure 4. Arachidonic acid (AA) inhibits the 4AP–evoked release of glutamate from rat cortical synaptosomes A: release of glutamate evoked by 1 mM 4AP with and without different concentrations of AA in the presence of 1.33 mM $CaCl_2$ B: Glutamate release induced by high concentrations of AA in polarized synaptosomes in the presence of 200 nM free $[Ca^{2+}]$ C: Glutamate release evoked by 30 mM KCl with and without AA in the presence of 1.33 mM $CaCl_2$ Each trace is the mean of three independent experiments [From Herrero et al., (43)]

the basal or the KCl–evoked release (34). Figure 3 shows that concentrations of PMA that cause the translocation of PKC do not increase the release of glutamate induced by 4AP. Higher concentrations of phorbol esters, however, significantly increase the glutamate release evoked by 4AP without effect on the KCl–induced release. The mechanism by which PKC potentiate glutamate exocytosis has been investigated by the parallel estimation of glutamate release, cytoplasmic free Ca^{2+} concentration, $[Ca^{2+}]_c$, and the plasma membrane potential (34). This study has shown that PKC activation does not alter the $[Ca^{2+}]_c$ of polarized synaptosomes but enhanced both the 4AP–evoked elevation in cytoplasmic Ca^{2+} and the 4AP–evoked depolarization. This potentiation of glutamate release by phorbol esters is consistent with a PKC–mediated inactivation of K^+ channels (35) which are likely to be the delayed rectifier involved in the termination of the action potential. Thus , since the Ca^{2+}–channels coupled to glutamate exocytosis do not undergo voltage inactivation (8), the inhibition of the delayed rectifier would prolong individual action potentials and hence enhance the entry of Ca^{2+} and the subsequent release of glutamate.

4. Arachidonic Acid Modulates Glutamate Exocytosis

In neurons arachidonic acid is released after the activation of the N–methyl–D–aspartate receptor (36). Recent work has suggested that this polyunsaturated fatty acid and its metabolites produced through the lipoxygenase pathways may constitute a novel class of second messenger (37) and evidence indicates that arachidonic acid participates in the modulation of glutamatergic synapses. In this context, it has been shown that arachidonic acid reduces the high affinity transport system for glutamate in synaptosomes (38,39). However, the presynaptic action of arachidonic acid in neurotransmitter release is less clear. Arachidonic acid has been reported to potentiate the K^+–evoked release of glutamate from hippocampal synaptosomes (40,41) but, endogenous free fatty acids released during the incubation of cerebrocortical synaptosomes inhibit the Ca^{2+}–dependent release of glutamate evoked with 4–aminopyridine but not with KCl (42). This inhibitory effect on glutamate release is also observed with exogenously added arachidonic acid, Fig 4. Low concentrations of arachidonic acid, (0.5–10 μM), inhibits the glutamate release induced by 4–aminopyridine, (Figure 4A) without effect on the slow Ca^{2+}–independent release of glutamate that occurs in the absence of depolarization, (Figure 4B). In contrast, arachidonic acid has no effect on the KCl–evoked release, (Figure 4C). The presynaptic inhibition by arachidonic acid is mediated by the modulation of K⁺–channels that control the duration of the action potentials. Thus the activation of K⁺ by arachidonic acid leads to a reduction in both the depolarization and the entry of Ca^{2+}, thereby resulting in a decrease in transmitter release (43). In agreement with this inhibitory effect by arachidonic acid on transmitter release, a specific inhibition of Ca^{2+}/calmodulin–dependent phosphorylation by arachidonic acid has also been observed in cerebrocortical synaptosomes on depolarization (44). Furthermore, a recent report in hippocampal synaptosomes has shown a reduction in the evoked release of glutamate in the presence of the 12–lipoxygenase product of arachidonic acid metabolism (45).

Protein kinase C is also activated by cis–fatty acids, such as oleic and arachidonic acid (46). Among the different subspecies of PKC , the γ–PKC is activated by arachidonic acid in the absence of diacylglycerol (47). However, all evidence obtained in nerve terminals indicates that the inhibition of glutamate release by low concentrations of arachidonic acid is not mediated by the activation of PKC, because of the insensitivity of the inhibitory effect to both the down regulation of PKC activity and to the presence of the protein kinases inhibitor staurosporine (48). In addition, glutamate exocytosis is inhibited by methylarachidonate which does not activate PKC (46).

Among the PKC subspecies present in cerebrocortical synaptosomes, arachidonic acid significantly activates the γ but not the α and β subspecies in the absence of phosphatidylserine and diacylglycerol (49). However, both the α and β subspecies highly

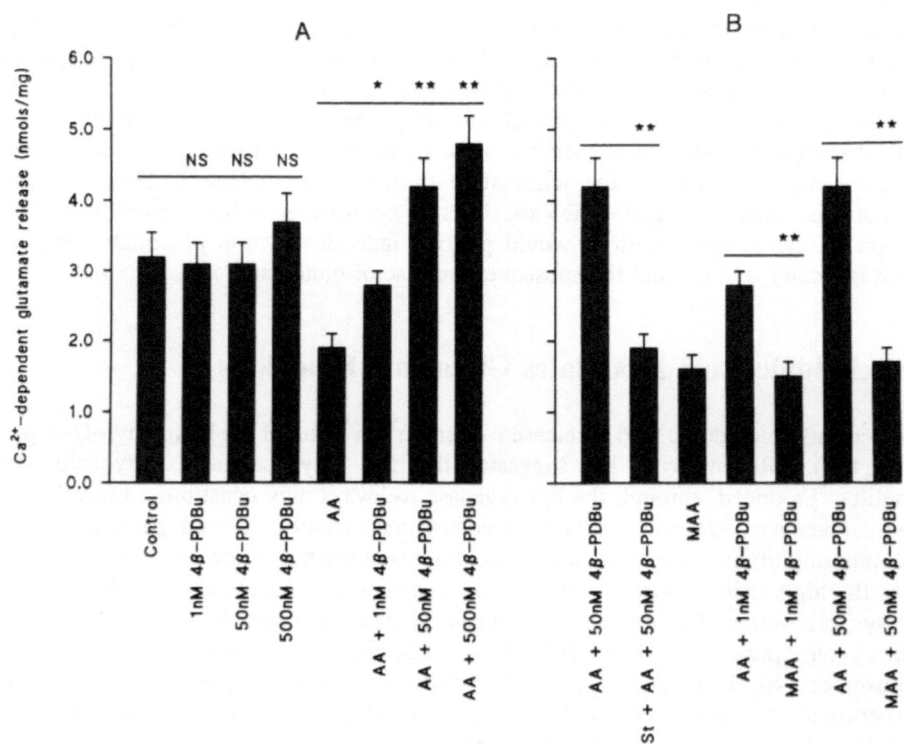

Figure 5. Low concentrations of 4β–phorbol dibutyrate (4β–PDBu) enhance glutamate exocytosis in the presence of arachidonic acid (AA) by a PKC–dependent mechanism. AA (2 μM) or methyl arachidonate (MAA; 2 μM) was added 1 min before depolarization of synaptosomes with 1 mM 4AP. The phorbol ester 4β–PDBu was added 30 s before depolarization. In experiments with staurosporine (St), synaptosomes were preincubated with the PKC inhibitor at 100 nM for 30 min. Results are means ± SEM (bars) values of three independent experiments. The values under the line were compared for significance of differences by the Student`s t test.: *p<0.05; **p<0.01; NS, not significant [From Herrero et al., (51)]

respond to arachidonic acid in the presence of diacylglycerol (50). In this context, in Figure 5A it is shown that concentrations of phorbol esters that do not significantly enhance the release of glutamate in the absence of arachidonic acid, greatly increase glutamate exocytosis in the presence of the fatty acid. The potentiation of glutamate release is consistent with the synergistic activation of the PKC subspecies present in nerve terminals by phorbol esters and arachidonic acid. In agreement with this, both the down regulation of PKC activity by pretreatment of synaptosomes with high concentrations of phorbol esters and the incubation with staurosporine prevent the potentiation of glutamate release, Figure 5B.

5. A Presynaptic Metabotropic Receptor for Glutamate Potentiates Glutamate Exocytosis

In the brain many neurotransmitters exert their actions by stimulating receptors linked to the phosphoinositide metabolism and by increasing the formation of the intracellular second messengers inositol trisphosphate, IP3, and diacylglycerol, DG. Although receptor activation of the phosphoinositide turnover has been shown to occur postsynaptically, there is some evidence regarding the presence of this mechanism for cellular signal transduction also at the presynaptic level. In rat brain synaptosomes, the stimulation of muscarinic receptors increase the formation of IP_3 (52) and the involvement of this second messenger in the mobilization of intracellular Ca^{2+} in synaptosomes has also been reported. (53). In addition, the activation of presynaptic protein kinase C potentiates the release of a number of neurotransmitters (30) including glutamate. (26,29,34).

Glutamate activates several types of receptors including a metabotropic receptor, sensitive to trans–1–amino–cyclopenthyl–1,3 dicarboxylate, ACPD, coupled to G protein(s) and linked

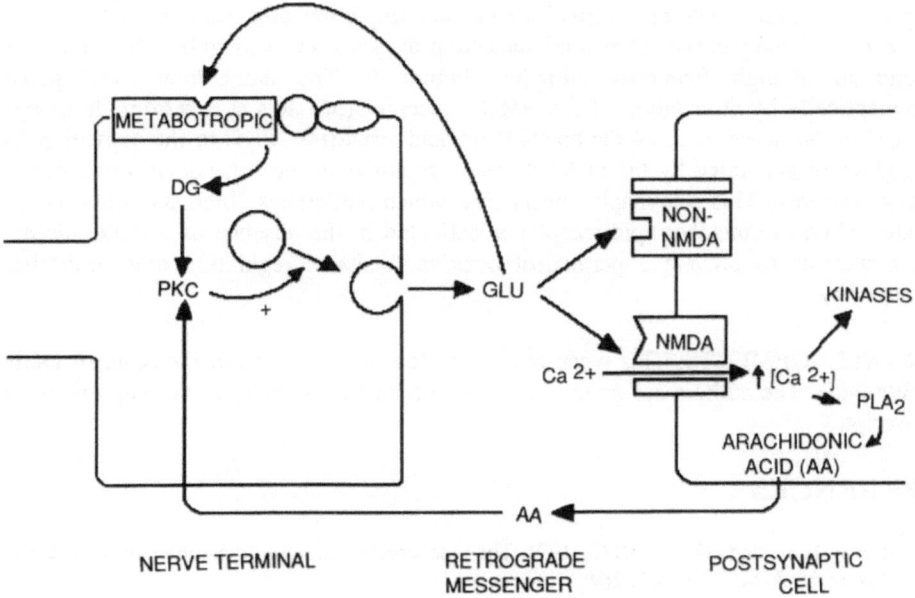

Figure 6. A PRESYNAPTIC METABOTROPIC RECEPTOR POTENTIATES THE RELEASE OF GLUTAMATE IN CEREBROCORTICAL NERVE TERMINAL. Glutamate (glu) released into the synaptic cleft activates non–N–methyl–D–aspartate (non–NMDA) receptors responsible for fast synaptic transmission, but fail to activate the NMDA receptor due to the voltage–dependent blockade by Mg^{2+}. Under repetitive stimulation, the NMDA receptor is activated and the Ca^{2+} entry through this receptor stimulates phospholipase A_2 (PLA_2), which generates arachidonic acid (AA). This "retrograde messenger" diffusses back to the presynaptic terminals where sensitizes the protein kinase C (PKC) to the activation by the diacylglycerol (DG) generated by the metabotropic receptor. The synergistic activation of PKC potentiates the presynaptic action potentials enhancing the release transmitter.

to the inositol phospholipid metabolism. (54). We have recently described the presence in cerebrocortical nerve terminals of a receptor for glutamate involved in the potentiation of glutamate exocytosis (55) that pharmacologically corresponds to the metabotropic type according to the following observations. First, the receptor is stimulated by low concentrations of the agonist ACPD. Second, the ACPD-mediated potentiation of glutamate release is insensitive to the ionotropic receptor antagonists 6-cyano-7-nitroquinoxaline, CNQX, and 2-amino-5-phosphonovaleric acid, AP_5, but sensitive to the relatively selective antagonist of the metabotropic receptor 2-amino-3-phosphonopropionate, $L-AP_3$. Finally, this receptor is coupled to a G protein sensitive to pertussis toxin.

The fact that the ACPD-induced potentiation of glutamate exocytosis is observed only in the presence of low concentrations of arachidonic acid (55) is consistent with the synergistic potentiation of PKC by the diacylglycerol generated by the metabotropic receptor and the fatty acid. Although the relatively specific selective agonist of the metabotropic receptor quisqualate releases Ca^{2+} from intracellular stores in cortical synaptosomes (56), the fact that the metabotropic receptor stimulation mimics the potentiation of glutamate release observed by the pharmacological activation of PKC with phorbol esters indicates that the effects by this receptor are mediated by the diacylglycerol/PKC branch of the phosphoinositide cascade rather than by the generation of IP_3.

The presence of a presynaptic autoreceptor in glutamatergic terminals involved in the potentiation of glutamate exocytosis is unusual since most autoreceptors prevent further release of neurotransmitter. Although such a presynaptic control for glutamate release could apparently result in neurotoxicity, it may be part of a positive feedback mechanism physiologically relevant in long term potentation, LTP. LTP is defined as a sustained increase in synaptic transmission that occurs in response to high frequency stimulation (57). It has been proposed that LTP in the hippocampus is initiated by the Ca^{2+} entry associated with the activation of NMDA receptors (58), followed by an increase in the efflux of glutamate (59). The presence in glutamatergic terminals of a metabotropic receptor that potentiates glutamate release only in the presence of arachidonic acid provides a mechanism by which the terminal respond to a high frequency stimulus, Figure 6. The arachidonic acid generated postsynaptically by stimulation of the NMDA receptor (36) acts as a retrograde messenger (60) and in the nerve terminal the arachidonic acid sensitizes PKC to the activation by the diacylglycerol generated by the metabotropic receptor with the subsequent enhancement in glutamate release. This presynaptic mechanism which potentiates glutamate release does not operate when the metabotropic receptor is activated in the absence of arachidonic acid in such a way as to prevent a permanent positive feedback regulation that could lead to neurotoxicity.

ACNOWLEDGEMENTS: This work was supported by a grant from the Spanish DGICYT (PM89/0045). The authors are grateful to Erik Lundin for his help in the preparation of the manuscript.

REFERENCES

1. McMahon, H.T., and Nicholls, D.G., 1991. The bioenergetics of neurotransmitter release. Biochim. Biophys Acta. **1059**:243-264.
2. Nicholls D.G., 1989 Release of glutamate and aminobutyric acid in isolated nerve terminals. J. Neurochem. **52**:331-3417.
3. Maycox, P.R., Hell, J.W., and Jahn, R., 1990. Amino acid neurotransmission; spotlight on synaptic vesicles. Trends in Neurosci. **13**:83-874
4. Nicholls, D.G., and Sihra, T.S., 1986. Synaptosomes possess an exocytotic pool of glutamate. Nature 321, 772-773.
5. Nicholls, D.G., and Attwell, D., 1990. The release and uptake of excitatory amino acids. Trends in Pharmacol. Sci. **11**:4683.

6. Nicholls, D.G., Sihra, T.S., and Sánchez–Prieto, J., 1987. Calcium dependent and independent release of glutamate from synaptosomes monitored by continuous fluorometry. J. Neurochem. 52:331–341.

7. Sánchez–Prieto, J., Sihra, T.S., and Nicholls, D.G., 1987. Characterization of the exocytotic release of glutamate from guinea pig cerebrocortical synaptosomes. J. Neurochem. 49:58–64.

8. McMahon H.T., and Nicholls, D.G., 1991. Transmitter glutamate release from isolated nerve terminals: evidence for biphasic release and triggering by localized Ca^{2+}. J. Neurochem. 56:86–94.

9. Sánchez–Prieto, J., Sihra, T.S., Evans, D., Ashton, A., Dolly, J.O., and Nicholls D.G., 1987. Botulinum toxin A blocks glutamate exocytosis from guinea pig cerebral cortical synaptosomes. Eur. J. Biochem. 165:675–681.

10. Verhage, M., McMahon H.T., Ghijsen W.E.J.M., Boomsma, F., Wiegant, V., and Nicholls D.G., 1991. Differential release of aminoacids, neuropeptides and catecholamines from nerve terminals. Neuron. 6:1–7.

11. Kauppinen, R.A., McMahon, H., and Nicholls D.G., 1988. Ca^{2+}–dependent and Ca^{2+}–independent glutamate release, energy status, and cytosolic free Ca^{2+} concentration in isolated nerve terminals following in vitro hypoglycaemia and anoxia. Neuroscience. 27:175–182.

12. Sánchez–Prieto, J., and González, M.P., 1988. Anoxia induces a large Ca^{2+}–independent release of glutamate in isolated nerve terminals. J. Neurochem. 50:1322–1324.

13. Rubio, I., Torres, M., Miras–Portugal, M.T., and Sánchez–Prieto, J., 1991. Ca^{2+}–independent release of glutamate during in vitro anoxia in isolated nerve terminals. J. Neurochem. 59:1159–1164

14. Choi, D.W., and Rothman S.M., 1990 The role of glutamate neurotoxicity in the hypoxic–ischemic neuronal death. Annu. Rev. Neurosci. 13:171–182.

15. Takai, Y., Kishimoto, A., Iwasa, Y., Kawahara, Y., Mori, T., and Nishizuka, Y., 1979. Calcium dependent activation of a multifunctional protein kinase by membrane phospholipids. J. Biol. Chem. 254:3692–3695.

16. Nishizuka, Y., 1986. Studies and perspectives on protein kinase C. Science 233:305–312. 17. Nishizuka, Y., 1988. The molecular heterogenity of protein kinase C and its implications for cellular regulation. Nature 339:661–665.

18. Akers, R.F., Lovinger, D.M., Colley, P.A., Linden D.J., and Routtenberg, a., 1986. Translocation of protein kinase C activity may mediate hIppocampal long term potentiation. Science 231:587–589.

19. Kikkawa, U., Kishimoto, A., Nishizuka, Y., 1989. The protein kinase C family: heterogenity and its implications. Annu. Rev. Biochem. 58:31–44.

20. Bell, R., 1986. Protein kinase C activation by diacylglycerol second messengers. Cell. 45:631–632.

21. Ganong, B., Loomis, C., Hannum, Y., and Bell, R., 1986. Specificity and mechanism of protein kinase C activation by sn–1,2–diacylglycerols. Proc. Natl. Acad. Sci. USA 83:1184–1187. 22. Wolf, M., Le Vine, M.III., May, W.S.Jr., Cuatrecasas, P., and Sahyoun, N., 1985. A model for intracellular translocation of protein kinase C involving synergism between Ca^{2+} and phorbol esters. Nature, 317:546–549.

23. Rodriguez–Peña, A., and Rozengurt, E., 1984. Disappearance of Ca^{2+}sensitive phospholipid–dependent protein kinase activity in phorbol ester–treated 3T3 cells. Biophys. Res. Commun. 120:1053–1059.

24. Girard, P.R., Mazzei, G.J., Wood, J.G., and Kuo, J.F., 1985. Polyclonal antibodies tophospholipids/Ca^{2+}–dependent protein kinase and inmunocytochemical localization of the enzyme. Proc. Natl. Acad. Sci. USA 82:3030–3034.

25. Shearman M.S., Shinomura, T., Oda, T., and Nishizuka., 1991. Synaptosomal protein kinase C subspecies: A., dynamic changes in the hippocampus and cerebellar cortex concomitant with synaptogenesis. J. Neurochem 56:1255–1262.

26. Díaz–Guerra, M.J.M., Sánchez–Prieto, J., Boscá, L., Pocock, J., Barrie, A., and Nicholls, D., Phorbol esters translocation and the potentiation of Ca^{2+}–dependent glutamate relase. (1988. Biochim. Biophys Acta. 970:157–165.

27. Oda, T., Shearman, M.S., and Nishizuka, Y., (1991. Synaptosomal protein kinase C subspecies: B., Down–regulation promoted by phorbol esters and its effect on evoked norepinephrine release. J. Neurochem. 56:1263–1269.

28. Takai, Y., Kishimoto, A., Inoue, M., and Nishizuka, Y., 1977. Studies on a cyclic nucleotide–independent protein kinase and its proenzyme in mammalian tissues. J. Biol. Chem. **252**:7603–7609.

29. Lynch, M.A., and Bliss, T.V.P., 1986. Long–Term Potentiation of synaptic transmission in the hippocampus of the rat: effect of calmodulin and oleoyl–acetyl–glycerol on the release of [^{3}H]–glutamate. Neuroscience Lett. **65**:171–176.

30. Nichols, R.A., Haycock, J. W., Wang, J.K.T., and Greengard, P., 1987. Phorbol ester enhancement of neurotransmitter release from the rat brain synaptosomes. J. Neurochem **48**:615–621.

31. Tanaka, C., Fujiwara, H., and Fujii, Y., 1986. Acetylcholine release from guinea pig caudate slices evoked by phorbol ester and calcium FEBS Lett. **195**:129–134.

32. Malenka, R.C., Ayoub, B.S., and Nicoll R.A., 1988. Phorbol esters enhance transmitter release in the rat hippocampal slices. Brain. Res. **403**:198–203.

33. Tibbs, G.R., Barrie, A.P., Van–Mieghem, F., McMahon, H.T., and Nicholls D.G., 1989. Repetitive action potentials in isolated nerve terminals in the presence of 4–aminopyridine: effects on cytosolic Ca^{2+} and glutamate release. J. Neurochem. **53**:1693–1699.

34. Barrie, A.P., Nicholls, D.G., Sánchez–Prieto, J., and Sihra, T.S., 1991. An ion channel locus for the protein kinase C potentiation of transmitter release from guinea pig cerebrocortical synaptosomes. J. Neurochem **57**:1398–1404.

35. Colby, K.A., and Blaustein, M.P., 1988. Inhibition of voltage gated K$^+$ channels in synaptosomes by sn–1,2–dioctanoylglycerol, an activator of protein kinase C. J., Neurosci. **8**:4685–4692.

36. Lazarewicz J.W., Wroblewski, J.T., and Costa E. J., 1990. N–methyl–D–aspartate–sensitive glutamate receptors induce calcium–mediated arachidonic release in primary cultures of cerebellar granule cells J. Neurochem. **55**:1875–1881.

37. Piomelli, D., and Greengard, P., 1991. Lipoxygenase metabolites of arachidonic acid in neuronal transmembrane signalling. Trends in Pharmacol. Sci. **11**:367–373.

38. Rhoads, D.E., Ockner, R.K., Peterson, N.A., and Raghutathy, E.., 1983., Modulation of membrane transport by free sodium–dependent amino acid uptake. Biochemistry. **22**:1965–1970.

39. Volterra, A., Trotti, D., Cassutti, P., Trombe, C., Salvaggio, A., Melgangi R.C., and Racagni, G., 1992. High sensitivity of glutamate uptake to extracellular free arachidonic acid levels in rat cerebrocortical synaptosomes and astrocytes. J. Neurochem. **59**:600–606.

40. Freeman, E., Terrian, D.M., Dorman, R.V., 1990. Presynaptic facilitation of glutamate release from isolated hippocampal mossy fiber nerve endings by arachidonic acid. Neurochem. Res. **15**:743–750.

41. Lynch, M.A. and Voss, K.L., 1990. Arachidonic acid increases inositol phospholipid metabolism and glutamate release in synaptosomes prepared from hippocampal tissue. J. Neurochem. **55**:215–221.

42. Herrero, I., Castro, E., Miras–Portugal, M.T., and Sánchez–Prieto, J., 1991. Glutamate exocytosis evoked by 4–aminopyridine is inhibited by free fatty acids released from rat cerebrocortical synaptosomes. Neurosci. Lett.**126**:41–44.

43. Herrero, I., Miras–Portugal, M.T., and Sánchez–Prieto, J., 1991. Inhibition of glutamate release by arachidonic acid in rat cerebrocortical synaptosomes. J. Neurochem. **57**:718–721.

44. Piomelli, D., Wang, J.K.T., Sihra, T.S., Nairn, A.C., Czernik, A.J., and Greengard, P., 1989. Inhibition of Ca^{2+}/calmodulin–dependent protein kinase II by arachidonic acid and its metabolites. Proc. Natl. Acad. Sci. USA. **86**:8550–8554.

45. Freeman, E.J., Damron, D.S., Terrian, D.M., and Dorman, R.V., 1991. 12–lipoxygenase products attenuate the glutamate release and Ca^{2+}–accumulation evoked by depolarization of hippocampal nerve endings. J. Neurochem **56**:1079–1082.

46. Murakami, K., and Routtenberg, A., 1985. Direct activation of purified protein kinase C by unsaturated fatty acids (oleate and arachidonate) in the absence of phospholipids and Ca^{2+} FEBS. Lett. **192**:189–193.

47. Shearman, M.S., Naor, Z., Sekiguchi, K., Kishimoto, A., and Nishizuka, Y., 1989. Selective activation of the subspecies of protein kinase C from bovine cerebellum by arachidonic acid and its lipoxygenase metabolites. FEBS Lett. **243**:177–182.

48. Herrero, I., Miras–Portugal, M.T., and Sánchez–Prieto, J., 1992. PKC–independent inhibition of glutamate exocytosis by arachidonic acid in rat cerebrocortical synaptosomes. FEBS Lett. **296**:317–319.

49. Shinomura, T., Asaoka, Y., Oka, M., Yoshida, K., and Nishizuka, Y., 1991. Synergistic action of diacylglycerol and unsaturated fatty acid for protein kinase C activation: its possible implications. Proc. Natl. Acad. Sci. USA **88**:5149–5153.

50. Shearman, M.S., Shinomura, T., Oda, T. and Nishizuka, Y.., 1991. Protein kinase C subspecies in adult rat hippocampal synaptosomes. Activation by diacylglycerol and arachidonic acid. FEBS Lett. **279**:261–264. 51. Herrero, I., Miras–Portugal, M.T., and Sánchez–Prieto. 1992. Activation of protein kinase C by phorbol esters and by arachidonic acid required for the optimal potentiation of glutamate exocytosis. J. Neurochem. **59**:1574–1577.

52. Audigier, S.M.P., Wang, J.K.T and Greengard, P., 1988. Membrane depolarization and carbamoylcholine stimulate phosphatidylinositol turnover in intact nerve terminals. Proc. Natl. Acad. Sci. USA **85**:2859–2863.

53. Gandhi C.R., and Ross, D.H., 1987. Inositol 1,4,5-trisphosphate induced mobilization of Ca^{2+} from rat brain synaptosomes. Neurochem. Res. **12**:67–72.

54. Schoepp, D., Bockaert, J., and Sladeczek, F., 1990. Pharmacological and functional characteristics of metabotropic excitatory amino acid receptors. Trends in Pharmacol. Sci. **11**:508–515.

55. Herrero, I., Miras–Portugal, M.T., and Sánchez–Prieto, J.., 1992. Positive feedback of glutamate exocytosis by metabotropic presynaptic receptor stimulation. Nature, **360**:163–166.

56. Adamson, P., Hajimohammadreza, I., Brammer, M.J., Campbell I.C., and Meldrum, B.S., 1990. Presynaptic glutamate/quisqualate receptors: effect on synaptosomal free calcium concentrations. J. Neurochem. **55**:1850–1854.

57. Bliss, T.V.P., and Lomo, T., 1973. Long lasting potentiation of synaptic transmission in the dentate of the anaesthetized rabbit following stimulation of the perforant path. J.Physiol. (Lond.) **232**:331–356.

58. Davies, S.N., Lester, R.A., Reyman, K.G., and Collingridge, G.L., 1989. Temporally distinct pre and postsynaptic mechanism maintain long–term potentiation. Nature. **338**:500–503.

59. Bliss, T.V.P., Douglas, R.M., Errington, M.L., and Lynch, M.A., 1986. Correlation between long–term potentiation and release of endogenous amino acids from dentate gyrus of anaesthetized rats. J. Physiol. (Lond.) **377**:391–408.

60. Williams, J.H., Errington, M.A., Lynch, M.A., and Bliss, T.V.P., 1989. Arachidonic acid induces long–term activity–dependent enhancement of synaptic transmission in the hippocampus. Nature **341**:739–741.

Controls of Cerebral Protein Breakdown

Abel Lajtha

1. Introduction

It is an honor and a pleasure to contribute to and participate in this prestigious course at the University Complutense, directed by the eminent scientist and teacher Dr. Santiago Grisolia. The topic of the course, dealing with the relationships of excess ammonia, liver pathology, and brain deficiency, indicates the interrelationships of the various organs in the body, and also indicates the need for us to identify the processes that are unique to an organ or are of special significance for it. One reason for selecting protein breakdown for discussion was its important relationship to ammonia, its products – amino acids. In general, only 0.1 to 1.0 percent of the amino acids are in the free amino acid pool as compared to the protein–bound forms; therefore, breakdown of only one percent of the proteins would increase the content in the free pool 2–10 fold. Therefore, protein breakdown could have an important influence on ammonia metabolism.

Whereas protein synthesis has occupied biological science for some time, resulting in great progress in understanding molecular biological processes and controls, protein breakdown studies were relatively neglected, perhaps because of the technical difficulties, or because the control of protein metabolism was thought to be exclusively in governing synthesis.

Now we know that although synthesis and breakdown are interrelated and need to be in balance, they represent two distinct and individually regulated processes. Furthermore, the biological role of breakdown, greatly expanded from the digestive one, is seen to extend to activation of enzymes from precursors in the nervous system, to formation of neuropeptides, and to many more. That is, the hydrolysis of proteins and peptides can represent the formation of functionally highly active compounds, not only their inactivation, and the process requires as much regulation as synthesis does.

Proteases in the brain are now at the center of interest, not only as a result of the above perceptions, but because of their proposed participation in processes of aging, of senility (amyloid protein formation), and of learning (long–term potentiation).

I want to illustrate the potential importance of proteolysis with an example, and the importance of studying processes, specifically in the brain, with another example.

If one looks among the various species at the biological parameters that parallel life span, a few correlations can be found – one is the rate of oxidative metabolism. Species with higher oxidative metabolism have correspondingly shorter life span. This finding, among others, led to the oxidation free radical hypothesis on the cause of aging – higher oxidative metabolism leads to faster production of free radicals, to more rapid aging, to shorter life. Another parameter is brain calpain (Ca–dependent neutral proteinase) content. The higher

---

Nathan S. Kline Institute for Psychiatric Research, Center for Neurochemistry, Orangeburg, NY 10962

Cirrhosis, Hyperammonemia, and Hepatic Encephalopathy,
Edited by S. Grisolia and V. Felipo, Plenum Press, New York, 1994

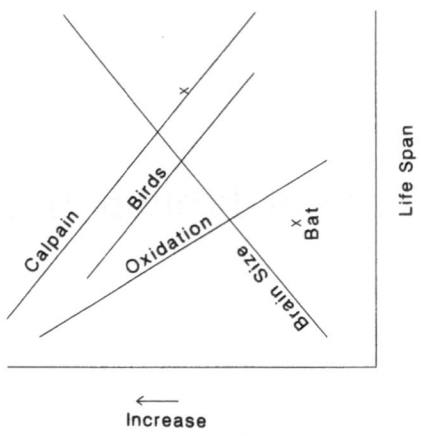

Figure 1

the calpain content the shorter the life span (1). This finding, among others, led to the proposal that proteinase activity participates in (or is responsible for) aging processes (2).

Figure 1 illustrates these points. As calpain content decreases in the species, brain life span is increased in vertebrates and in birds in parallel fashion, as with oxidative metabolism. Brain size and life span are directly proportional; the larger the brain, the longer the life span.

The example that the controls of brain protein breakdown are different from those of the rest of the organism can, unfortunately, be seen often in malnourished people. An extreme example is the victims of concentration camps, who exhibited loss of much of their total body protein, without significant loss of brain proteins. We examined the effects of malnutrition on brain protein metabolism (3); while in muscle (used for supplying amino acids) protein synthesis was decreased in malnutrition, protein breakdown was increased – leading to net loss of protein; in brain, protein synthesis was decreased, but breakdown was also decreased (under mild conditions to the same extent), resulting in no net loss of proteins. In extreme cases, this process can become unbalanced, especially in the developing nervous system, and some protein deficiency in the adult brain can be observed (Table 1).

This illustrates how brain differs from muscle, in that whereas breakdown is inhibited in brain it is stimulated in muscle, where the combination of inhibition of synthesis and stimulation of breakdown leads to significant loss of muscle protein content. The loss of protein in brain in severe malnutrition is of great importance, since it can happen to malnourished children in many parts of the world. Biological defence mechanisms are sometimes not adequate.

Table 1. Protein metabolism in malnutrition

Tissue	Age	Synthesis	Breakdown	Net Change
Muscle	Adult	–	+	– –
Muscle	Young	– –	+ +	– – – –
Brain	Adult	–	–	0
Brain	Young	– –	–	–

2. Rates of Brain Protein Synthesis

There is now fairly good agreement of a number of different laboratories using different methods, such as administering the labeled precursor amino acid used for incorporation measurements as a tracer dose, as a large flooding dose, in constant infusion, as a pellet (or in form of labeled glucose, giving rise to amino acids). All indicated an average rate of incorporation in the young brain of about 2% per hour, and in the adult brain of about 0.7% per hour (4–8) (Table 2). The values converted into half–lives (the time it takes for half the proteins to be replaced (broken down and resynthesized), indicated two pools in the adult brain (about 4% of the total with a half-life of 7 h and 96% with a 10–day half–life), and a single pool in the young (2–day half–life). These values indicate a surprisingly rapid metabolism of proteins in the brain (12,13), an organ with little or no regeneration, but one storing memory (Table 3). The metabolic rate in liver is considerably higher than that in brain; in muscle it is lower.

It seems that this metabolic activity – constant breakdown and resynthesis – is characteristic of most cerebral proteins. In a study on the stability of cerebral proteins we labeled all brain proteins: we fed the animals a diet containing labeled amino acids of constant specific activity before pregnancy, during pregnancy and lactation, and then the offspring during growth, so that the experimental animals were exposed only to labeled food throughout their entire growth. Any protein that was stable should have retained its radioactivity when the diet was replaced with one containing no labeled aminoacid. With time, however, most of the label was replaced in the brain, indicating that most of its proteins undergo continuous metabolism (9).

Table 2. Initial synthesis rates for whole brain proteins

| Method | Percent per hour | |
	Adult	Young
Pulse	0.61	2.0
		2.2
Infusion	0.68	2.0
		0.55
		0.87
Massive dose	0.62	2.1
	0.78	2.0
Multiple injection	0.80	
Pellet implantation	0.65	
Average	0.69	2.0

Table 3. Tissue protein turnover rates

Organ	Pool %	Half-Life hours
Brain, adult	4	7
	96	230
Brain, young	100	48
Liver	100	26
Kidney	40	18
	60	63

The release of label (Table 4) was slower than the incorporation. This is due to the re–use of the amino acid liberated by protein breakdown for synthesis.

These examples thus indicate that there is an active and extensive process of protein breakdown that has to be under specific control, and must be in balance with protein synthesis.

So far, the discussion has covered studies that were done on living animals. There is very ample literature reporting important findings using isolated (in vitro) systems. Although

Table 4. Instability of cerebral proteins

Time after change (days)	Percent label remaining
0	100
30	30
150	2

for analyzing complex mechanisms such as protein synthesis, studies with isolated systems are essential, for metabolic rates over time, the experiments can only be done in vivo. This is illustrated by comparing incorporation of amino acids in vivo and in brain slices. The two measurements, in vivo and in vitro, are fairly close in the immature tissue, but incorporation in slices from adult brain represents only a small fractions of the in vivo rates (Table 5). It is of interest that this difference between the rates in vivo and in slices is less in the adult pituitary and pineal, just as the developmental decrease in rates in vivo is also less (6).

Table 5. Comparison of in vitro to in vivo synthesis rates in rat brain during developement

	3 days	10 days	23 days	Adult
	Protein synthesis in vivo percent per hour			
Cerebrum	2.0	1.6	0.93	0.62
Cerebellum	2.9	2.5	1.1	0.70
Pituitary	2.7	2.6	1.8	1.4
Pineal	3.0	2.8	2.6	2.2
	Incorporation in vitro as per cent of that in vivo			
Cerebrum	79	75	15	11
Cerebellum	79	90	30	6
Pituitary	71	69	76	76
Pineal	68	70	70	78

3. Some Properties of Cerebral Protein Breakdown

We still know little about the mechanisms involved in protein catabolism. A large number of enzymes have been characterized: several classes of proteinases and peptidases and their substrate specificity were examined. In spite of these efforts, the particular enzyme or enzymes that metabolize each protein substrate is not known. It is possible that considerable heterogeneity exists and that the breakdown of a protein is dependent on its location. A possible example is shown in Table 6 (10): membrane–bound tubulin from brain is resistant to cerebral cathepsin D action, while cytosolic tubulin is rapidly hydrolyzed. The membrane–bound form is not resistant because protein is made inaccessible by the membrane, since the membrane is hydrolyzed.

It is generally thought that cathepsin D acts in lysomal proteolysis only in highly acidic environment. Its activity may also depend on the substrate. We (11) and others found that the pH dependence of cathepsin D is different for different substrates – more acidic for hemoglobin (a usual substrate for cathepsin studies) than for tubulin, for example. Since cathepsin D is present at high levels, even fractional activity not close to its pH optimum can be a physiologically important function. Thus, many proteases are likely to participate in protein turnover, with partial proteolysis resulting in the formation of active compounds and further proteolysis in inactivation. The peptidase activity that metabolizes protease products

Table 6. Tubulin breakdown by cathepsin D and thrombin

Enzyme	Substrate	% Breakdown	
		2 h	4 h
Cathepsin D	(pH 5.8) cytosol tubulin	54	79
	membrane tubulin	0	7
Thrombin	(pH 8.6) cytosol tubulin	65	80
		23	46
	membrane tubulin	25	39
		18	32

further is considerably higher than the protease activity, and very little of the peptide intermediates can be found in the brain.

Clearly, protein synthesis and breakdown have to be regulated in parallel fashion: more rapidly formed proteins must also be more rapidly catabolized so that the protein composition is not altered under most circumstances. It is not clear how such parallel regulation works.

4. Alterations of Protein Breakdown

The change in protein synthesis during development was already discussed. We found that breakdown rates in vivo also undergo developmental changes (14), and are higher in young brain (Fig. 2).

The fact that during growth synthesis rates are above breakdown rates accounts for the net deposition of proteins, which stops when synthesis and breakdown become equal. The more rapid breakdown during the most active growth phase is somewhat surprising, and is possibly explained by the high rate of cell death at this stage – cells not developing the proper connections are eliminated.

There are changes in proteases during aging. These changes are puzzling because the in vitro and in vivo changes are not parallel. We found a large increase in cathepsin D content in the aging brain (15). In most regions the level increased two–fold or more (Table 7).

Such an increase was indicated before, in a study finding greater breakdown of microtubule–associated proteins (16). Immunoassay of cathepsin D content indicates a greater content of the inactive enzyme form in aging (17).

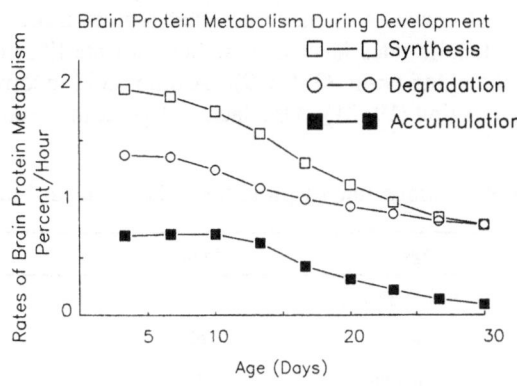

Figure 2

Calpain (Ca–dependent neutral protease) activity also increases in the brain. It has to be emphasized that, as with cathepsin D, measurement of enzyme activity must be done on isolated systems, if possible under optimal conditions; in the case of calpains, after removal of the endogenous inhibitor. Thus, such measurements indicate enzyme content rather than in vivo enzyme activity. Since, as will be discussed further on, only a fraction of the enzyme content of the tissue is needed for the catabolic portion of in vivo protein turnover, cathepsins and calpains must be present partially in an inactive form or in a form that is much less active than that measured in vitro.

Table 7. Changes in cathepsin D activity in aging rat brain.

Regions	% Increase
Cortex	106
Cerebellum	151
Pons–medulla	145
Striatum	142
Hippocampus	114
Spinal cord	93

Specific activity in 24–month–old rat brain compared to 3–month–old (nmol tyrosine equivalent per mg protein per hr).

The increase in calpain activity was different in different brain regions (Table 8). We also found some variations in regional distribution of activity that were dependent on the substrate used (18). This variation indicates a change in the properties of the enzyme

Table 8. Age–related changes of CANP II activity in various brain regions.

Region	Activity at 24 months as percent of that at 3 months		
	Casein	Desmin	Actin
Cortex	172	91	128
Cerebellum	142	96	122
Pons–medulla	101	100	100
Striatum	163	182	195
Hippocampus	117	133	99
Hypothalamus	116	146	140
Spinal cord	105		

(substrate specificity?) with aging. The increase in enzyme content may be a result of changes in the properties of enzymes or substrates that require more enzyme. It does not indicate an increase in metabolic rates in vivo – studies indicate little change or a decrease of protein turnover in the aging brain (Table 9). Amino acid incorporation into proteins shows slight alteration with aging (19–21). One factor of great importance in in vivo protein

Table 9. Average rate of protein synthesis in rat brain

	Age	%/hr	Activity %
Young	0–5 days	2.1	330
Adult	6 months	0.63	100
Old	26 months	0.49	78

metabolic rates is body temperature. We found a decrease of about 6% per each ^{0}C lowering of body temperature (22). This is of importance in several aspects. A number of reports of drug effects on brain protein metabolism represent not direct drug effects on brain, but effects on body temperature. Fish live in water that differs 15^{0}C or more between summer and winter, therefore their body temperature is 15^{0}C lower in winter. This would mean 90–95% drop in brain protein turnover rates under winter conditions (23). It is of interest that not all processes in the brain show such high temperature dependence. Amino acid transport, for example, changes less than 1% per 1C^0. This may be useful in therapeutic procedures, where processes such as protein pathological degradation (due to brain injury) or formation can be strongly inhibited without a comparable degree of interference with the metabolic supply (and product removal) in the brain.

The body temperature seems to influence cerebral metabolic rates in all species (24), but the rate in cold–blooded ones may be below that in warm–blooded ones even at 37^{0}C (Table 10).

Table 10. Brain protein synthesis rates in various species.

Species	nmol Valine mg/protein	% Incorporation/h 22^{0}C	37^{0}C
Goldfish	453	0.23	
Bullfrog	447	0.18	
Lizard	458	0.13	
			0.27
Chicken	468		0.66
Mouse	465		0.65

5. Stimulation of Calpain Activity

The early work attempting to regulate protease activity focused on inhibitors. These compounds, which can very efficiently inhibit a class of proteases, were of great help in enzyme studies, but we still do not have highly specific inhibitors. Attention has only recently turned to factors that stimulate their activity. The presence of an endogenous proteinacious activator was reported (25), but it has not been fully characterized yet. In human neutrophils isovalerylcarnitine was reported to increase Ca affinity of calpains, thereby activating the enzyme at low Ca levels (26). This is of importance since the Ca requirements of calpains are well above physiological Ca levels. More recently, polyamines were found to activate calpains in crude but not in pure preparations (27). This finding was interpreted

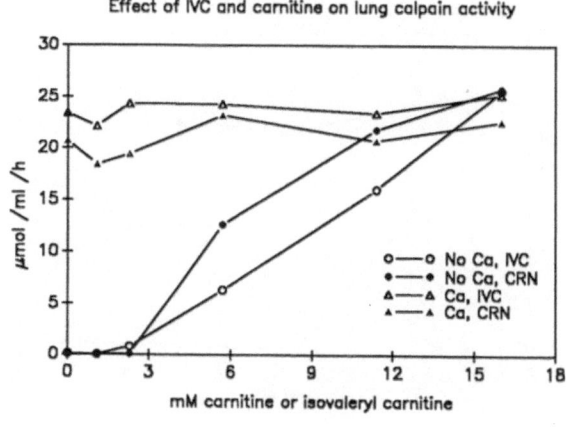

Figure 3

to mean that polyamines enhance the effects of an endogenous activator, which is present in crude but not in pure preparations.

Our studies indicated that isovalerylcarnitine has different effects on the calpains from different organs. While in neutrophils it does not replace Ca (28), in crude lung extracts it fully activates the enzyme even in the absence of Ca (Fig. 3).

Isovalerylcarnitine, or carnitine, in lung preparations can replace Ca, but it does not increase enzyme activity beyond to that in the presence of Ca alone.

In brain, isovalerylcarnitine or carnitine not only replaces the Ca requirement, but increases activity to a level severalfold higher than that in the presence of Ca alone (Fig. 4).

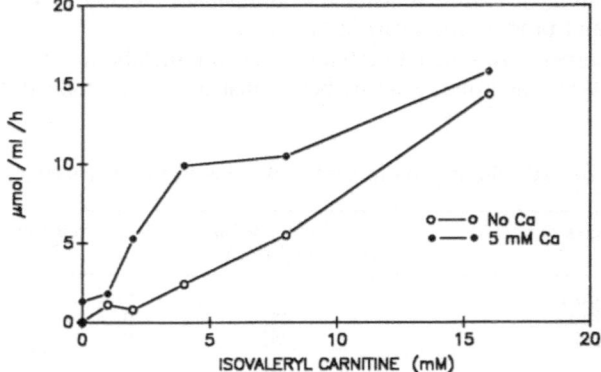

Figure 4. Effect of IVC on calpain activity in rat brain supernatant in the presence or absence of calcium.

This activation is not a liberation of some bound Ca stores, since activity in the presence of Ca is fully inhibited by EGTA, but not in the presence of carnitine (Fig. 5). Carnitine and isovalerylcarnitine, like polyamines, do not react with the enzyme directly, since they have no effect on the purified enzyme.

The Ca dependence curve of brain calpain in a pure enzyme preparation is not affected by isovalerylcarnitine (Fig. 6). The activation does not seem merely to produce more enzymes, for example, from inactive precursor via an autolytic process, because leupeptin, which strongly inhibits activity, has no effects in the presence of carnitine (Fig. 7). In purified brain preparations leupeptin strongly inhibits or abolishes calpain activity even in the presence of carnitine.

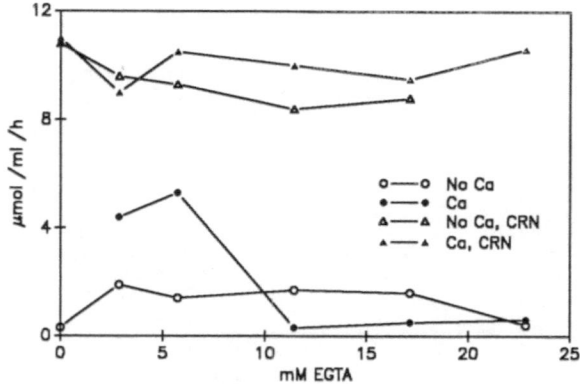

Figure 5. Carnitine and EGTA effects on crude brain calpain.

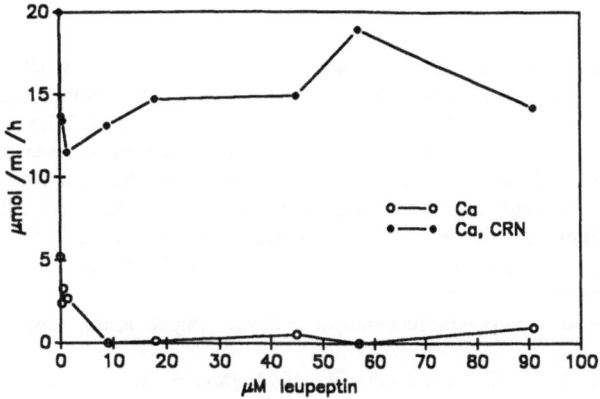

Figure 6. Effect of carnitine on leupeptin inhibition of crude brain calpain.

We do not know what the mechanism of activation is in the crude calpain preparations using carnitine. The effect seems to be complex, since in lung extracts carnitine merely replaces Ca, while in brain extracts it stimulates activity severalfold. Several activation mechanisms have been described, such as by membrane association (29), by phorbol ester (30), and by endogenous activators isolated from platelets (31) and muscle (32).

Because our recent work centered on stimulation of calpain activity, this was discussed and illustrated in more detail. Clearly, control requires both stimulation and inhibition of activity. As already mentioned, the enzyme (protease and peptidase) content of the brain is severalfold higher than needed for physiological activity. It is likely that in the living tissue some of the enzyme is occluded or not fully active, and therefore is not under optimal conditions. The capacity is so great that if the enzyme were fully active it could degrade the total brain protein content within hours.

We still know very little about how the possible inhibitors and activators would act, and under what conditions. It is likely that more knowledge about them not only will have importance in our understanding of the basic biological mechanisms in brain function, but also will help in therapeutic processes; for example, in Alzheimer's disease, inhibition of the formation of the toxic amyloid proteins, or after their formation, the stimulation of their breakdown. Since protein metabolism is involved in many other brain functions, among them long–term potentiation and memory formation, the findings may be of importance in influencing normal functions as well.

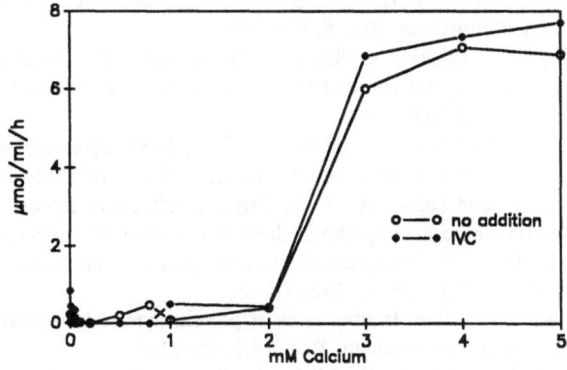

Figure 7. Calcium dependence and IVC effect with purified brain calpain.

REFERENCES

1. Baudry, M., DuBrin, R., Beasley, L., Leon, M., and Lynch, G., 1986, Low levels of calpain activity inchiroptera brain: Implications for mechanisms of agin, Neurobiol. Aging 7:255–258.
2. Lynch, G., Larson, J., and Baudry, M., 1986, Proteases, neuronal stability, and brain aging: an hypothesis. In: Treatment Development Strategies for Alzheimer's Disease, (Crook, T., Bartus, R.T., Ferris, S., and Gershon, S. eds.). Mark Powley Assoc., Inc., Madison, CT, pp. 119–149.
3. Banay–Schwartz, M., Giuffrida, A. M., DeGuzman, T., Sershen, H., and Lajtha, A., 1979, Effect of undernutrition on cerebral protein metabolism, Exp. Neurol. 65:157–168.
4. Austin, L., Lowry, O. H., Brown, J. G., and Carter, J. G., 1972, The turnover of protein in discrete areas of rat brain, J. Biochem. 126:351–359.
5. Oja, S. S., 1967, Studies on protein metabolism in developing rat brain, Ann. Acad. Sci. Fenn. A5. 131:1–81.
6. Dunlop, D. S., Van Elden, W., and Lajtha, A., 1977, Developmental effects on protein synthesis rates in regions of the CNS in vivo and in vitro, J. Neurochem. 29:939–945.
7. Seta, K., Sansur, M., and Lajtha, A., 1973, The rate of incorporation of amino acids into brain proteins during infusion in the rat, Biochim. Biophys. Acta. 294:472–480.
8. Dunlop, D. S., Van Elden, W., and Lajtha, A., 1975, A method for measuring brainprotein synthesis rates in young and adult rats, J. Neurochem. 24:337–344.
9. Lajtha, A., and Toth, J., 1966, Instability of cerebral proteins, Biochem. Biophys. Res. Commun. 23:294–298.
10. Bracco, F., Banay–Schwartz, M., DeGuzman, T., and Lajtha, A., 1982, Membrane–bound tubulin: resistance to cathepsin D and susceptibility to thrombin, Neurochem. Int. 4:501–511.
11. Bracco, F., Banay–Schwartz, M., DeGuzman, T., and Lajtha, A., 1982, Brain tubulin breakdown by cerebral cathepsin, D. Neurochem. Int. 4:541–549.
12. Lajtha, A., Dunlop, D., Patlak, C., and Toth, J., 1979, Compartments of protein metabolism in the developing brain, Biochim. Biophys. Acta. 561:491–501.
13. Lajtha, A., Latzkovits, L., and Toth, J., 1976, Comparison of turnover rates of proteins of the brain, liver, and kidney in mouse in vivo following long–term labeling, Biochim. Biophys. Acta. 425:511–520.
14. Dunlop, D. S., Van Elden, W., and Lajtha, A., 1978, Protein degradation rates in regions of the central nervous system in vivo during development, J. Biochem. 170:637–642.
15. Kenessey, A., Banay–Schwartz, M., DeGuzman, T., and Lajtha, A., 1989, Increase in cathepsin D activity in rat brain in aging, J. Neurosci. Res. 23:454–456.
16. Matus, A., and Green, G. D. J., 1987, Age–related increase in a cathepsin D like protease that degrades brain microtubule–associated protein, Biochemistry. 26:8083–8086.
17. Wiederanders, B., and Oelke, B., 1984, Accumulation of inactive cathepsin D in old rats, Mech. Age. Dev. 24:265–271.
18. Kenessey, A., Banay–Schwartz, M., DeGuzman, T., and Lajtha, A., 1990, Calpain II activity and calpastatin content in brain regions of 3– and 24–month–old rats, Neurochem. Res. 15:243–249.
19. Fando, J. L., Slainas, M., and Wasterlain, C. G., 1990, Age–dependent changes in brain protein synthesis in the rat, Neurochem. Res. 5:373–383.
20. Avola, R., Condorelli, D. F., Ragusa, N., Renis, M., Alberghina, M., Giuffrida Stella, A.M., and Lajtha, A., 1988, Protein synthesis rates in rat brain regions and subcellular fractions during aging, Neurochem. aes. 13:337–342.
21. Ingvar, M. C., Maeder, P., Sokoloff, L., and Smith, C. B., 1985, Effects of ageing on local rates of cerebral protein synthesis in Sprague–Dawley rats, Brain 108:155–170.
22. Sayegh, J. F., Sershen, H., and Lajtha, A., 1992, Different effects of hypothermia on amino acid incorporation and on amino acid uptake in the brain in vivo, Neurochem. Res. 17:553–557.
23. Lajtha, A., and Sershen, H., 1975, Changes in the rates of protein synthesis in the brain of goldfish at various temperatures, Life Sci. 17:1861–1868.
24. Sayegh, J. D., and Lajtha, A., 1989, In vivo rates of protein synthesis in brain, muscle, and liver of five vertebrate species, Neurochem. Res. 14:1165–1168.
25. DeMartino, G. N., and Blumenthal, D. K., 1982, Identification and partial purification of a factor that stimulates calcium–dependent proteases, Biochemistry. 21:4297–4303.
26. Pontremoli, S., Melloni, E., Michetti, M., Sparatore, B., Salamino, F., Siliprandi, N., and Horecker,

B. L., 1987, Isovalerylcarnitine is a specific activator of calpain of human neutrophils, Biochem. Biophys. Res. Commun. **148**:1189–1195.

27. Najm, I., Vanderklish, P., Etebari, A., Lynch, G., and Baudry, M., 1991, Complex interactions between polyamines and calpain–mediated proteolysis in rat brain. J. Neurochem. **57**:1151–1158.

28. Pontremoli, S., Melloni, E., Viotti, P. L., Michetti, M., Di Lisa, F., and Siliprandi, N., 1990, Isovalerylcarnitine is a specific activator of the high calcium requiring calpain forms, Biochem. Biophys. Res. Commun. **167**:373–380.

29. Pontremoli, S., Sparatore, B., Salamino, F., Michetti, M., Sacco, O., and Melloni, E., 1985, Reversible activation of human neutrophil calpain promoted by interaction with plasma membranes, Biochem. Int. **11**:35–44.

30. Pontremoli, S., Melloni, E., Salamino, F., Patrone, M., Michetti, M., and Horecker, B. L., 1989, Activation of neutrophil calpain following its translocation to the plasma membrane induced by phorbol ester or fMet–Leu–Phe, Biochem. Biophys. **160**:737–743.

31. Shiba, E., Ariyoshi, H., Yano, Y., Kawasaki, T., Sakon, M., Kambayashi, J., and Mori, T., 1992, Purification and characterization of a calpain activator from human platelets, Biochem. Biophys. Res. Commun. **182**:461–465.

32. Pontremoli, S., Viotti, P. L., Michetti, M., Sparatore, B., Salamino, F., and Melloni, E., 1990, Identification of an endogenous activator of calpain in rat skeletal muscle, Biochem. Biophys. Res. Commun. **171**:569–574.

Two Different Families of NMDA Receptors in Mammalian Brain: Physiological Function and Role in Neuronal Development and Degeneration

Elias K. Michaelis

1. Introduction

The dicarboxylic amino acids L–glutamic and L–aspartic acid are the most prevalent excitatory neurotransmitters in both the vertebrate and invertebrate nervous systems (1–3) The activity of these excitatory systems is intimately involved in the processing of incoming information into the spinal cord and brain, the initiation of learning processes, and the formation of memory (2–4). Glutamate neurotransmission involves most brain regions and subserves several important physiological functions. It is apparently the vehicle for rapid excitatory neurotransmission in areas such as the cerebral and cerebellar cortex, the limbic system, thalamus, basal ganglia, brain stem and spinal cord (5). In addition to the rapid synaptic transmission of neuronal pulses by the glutamate neurotransmitter system, this excitatory system also employs receptors that bring about more prolonged excitation of neurons either through activation of entry of extracellular calcium (Ca^{2+}) into the neuronal space or through the release of intracellular Ca^{2+} from internal stores (6–9). The events initiated by the activation of special subpopulations of glutamate receptors are thought to be instrumental in the long–term alterations of synaptic efficacy associated with such physiologic phenomena as long term potentiation of synaptic activity (2–4,10).

A prolonged activation of some receptors by the agonist L–glutamate brings about a much enhanced entry of Ca^{2+} into neurons, and this Ca^{2+} ion flux may function as the triggering stimulus for the initiation of neurodegenerative events (11,12). The scenario has been demonstrated quite clearly in pathological states such as hypoxia, hypoglycemia, strokes and seizures (13–17). There is also initial evidence that in states of induced hyperammonemia in experimental animals, some of the same receptors for glutamate as those involved in the conditions described above are hyperactivated and may be an intermediate step in the appearance of neurodegeneration associated with this condition (18,19).

Because of the multitude of pathological states of the central nervous system that may involve glutamate receptor hyperactivity as an initiating event in neurodegeneration, the definition and characterization of glutamate receptors has received great degree of scrutiny within the last decade. These receptors or the events initiated following receptor activation may form important targets for therapeutic intervention in acute or chronic neurodegenerative conditions. The study of these receptors was initially very difficult and the results obtained

Department of Pharmacology and Toxicology and the Center for Neurobiology and Immunology Research, University of Kansas, Lawrence, KS, 66045, USA

Cirrhosis, Hyperammonemia, and Hepatic Encephalopathy,
Edited by S. Grisolia and V. Felipo, Plenum Press, New York, 1994

were at times confusing. With the development of more selective pharmacological tools it has become apparent that this key neurotransmitter system in brain and spinal cord uses at least four variations of receptors to produce its physiological activity and that some of these receptors have very unusual properties such as high conductance of Ca^{2+} and voltage-dependent inactivation by divalent cations such as magnesium (Mg^{2+}).

Activation of neuronal receptors by L–glutamate leads to one of two types of responses, either stimulation of phospholipase C (8,20) or increases in neuronal membrane Na^+, K^+ and Ca^{2+} permeability and neuronal excitation (e.g., 6,7,21,22). The receptors that are linked to ion channels are subclassified into three major types: those activated by alpha–amino–3–hydroxy–5–methyl–4–isoxazole–propionic acid (AMPA), those that respond to kainic acid, and those that are sensitive to N–methyl–D–aspartate (NMDA) (23). The most intriguing of these receptors is the one that is activated by NMDA, since it is this receptor that has a high conductance of Ca^{2+}, is regulated by Mg^{2+} ions, is involved in the phenomena of long–term potentiation and long–term depression of synaptic activity, is developmentally regulated with respect to its expression, it affects neuronal survival or synaptogenesis, and is apparently responsible for the initiation of neurodegeneration following episodes of hypoxia, hypoglycemia, strokes, seizures, or hyperammonemia (10,15–18,24–27). As a consequence of the assignment of these crucial functions to this one subtype of glutamate receptors, it is not surprising that so much attention has been focused on the molecular, physiological and pharmacological characteristics of the NMDA receptor.

Most current knowledge about the physiologic activity of the NMDA receptors was accumulated through the use of competitive antagonists of the receptors such as 2–amino–7–phosphonoheptanoic acid (2–AP7) (28,29), 2–amino–5–phosphonopentanoic acid (2–AP5) (23), and carboxypiperazinylpropylphosphonic acid (CPP) (30,31). The complexity of regulation of the NMDA receptor was recently revealed by the observation that glycine, D–serine, and polyamines, such as spermine, allosterically potentiate activation of the receptor by agonists (32,33). Channel conductance of NMDA receptors is blocked in a voltage-dependent manner by physiologic concentrations of Mg^{2+} (6,34,35). Channel conductance of the NMDA receptors is also inhibited through what has been characterized as "open–channel blockade" by the arylcyclohexylamines PCP and ketamine, the benzomorphans NANM and cyclazocine, the substituted dioxolanes etoxadrol and dexoxadrol, the dibenzylcyclohepteneimine MK–801, and the morphinan derivatives dextromethorphan and dextrorphan (36–42). There is strong evidence that these agents bind to the open state of the activated NMDA receptor–ion channel complex to produce the observed channel block (39,43,44).

2. Definition of Ligand Binding Sites of NMDA Receptors and Isolation and Reconstitution of Glutamate/NMDA Receptors

When the ligand binding sites associated with the glutamate/NMDA receptors were labeled *in situ* in rat brain by agents that are selective for the agonist, competitive antagonist, co-activator, and non–competitive antagonist sites of the receptor complex, several apparent disparities were revealed in terms of the density of sites labeled by each of these agents in various brain regions. Monaghan and colleagues (45–47) have shown that NMDA–sensitive L–[^{3}H]glutamate binding sites had a substantially different distribution in brain from the sites labeled by [^{3}H]CPP, a competitive inhibitor of NMDA activation of the receptors and of L–[^{3}H]glutamate binding (48). In addition, they showed that there is marked regional variation in the affinity of the receptors for competitive as well as noncompetitive NMDA receptor inhibitors. An initial explanation offered by these investigators was that there may be agonist–preferring and antagonist–preferring states of the receptor. A more likely explanation that may account for this marked variation is that there exist different forms of the NMDA receptor. Other investigators have also reported low affinity binding of non–competitive ligands such as [^{3}H]MK–801 to NMDA receptors expressed in cerebellar neurons as compared to receptors in the frontal regions of the brain (49). These observations may be viewed as further evidence that NMDA receptors are a heterogeneous macromolecular complex that exhibits substantial variability when it is expressed by neurons in different locations of the central nervous system.

Inactivation by ionizing radiation of NMDA–sensitive L-[^{3}H]glutamate binding sites, strychnine–insensitive [^{3}H]glycine sites, [^{3}H]MK–801 sites and [^{3}H]TCP binding sites has been used to estimate the M_r for each entity that binds these ligands: 121,000, 115,000, 128,000 and 118,000 for the glutamate, glycine, MK–801 and TCP binding entities, respectively (50,51). Sonders and colleagues (52) developed an azido derivative of the ion channel blocker MK–801. Following photoactivation of this derivative in the presence of neuronal membranes from previously frozen guinea pig brain, a single polypeptide was labeled of estimated molecular size of 120 kDa. These findings are consistent with a structure for the NMDA receptors as consisting of a homomeric complex of proteins of ~120 kDa molecular size. On the other hand, Honoré and colleagues (50) demonstrated that the kinetics of radiation–induced inactivation of [^{3}H]CPP binding performed under identical conditions as those used for measuring the effects on glutamate, glycine and TCP sites, led to the estimation of a molecular size equal to 209 kDa for the CPP binding sites. These investigators proposed a model for the receptor which includes two subunits, one of M_r ~120,000 which has the glutamate, glycine and TCP binding sites and the other of M_r ~80,000 that has the CPP binding sites. Since CPP behaves as a competitive antagonist in physiological and ligand binding studies, an implicit assumption made by these investigators was that the CPP binding site overlaps with the glutamate/NMDA binding site.

The model proposed by Honoré and colleagues implies a heteromeric structure of the NMDA receptor. A heteromeric structure for the NMDA receptor is also implied by Haring *et al.* (53), who reported that the photoactivatable derivative [^{3}H]azidoPCP labeled several broad bands of proteins in synaptic membranes, the major ones being the 90, 62, 49, 40, and 33 kDa proteins. In a more recent report from the same laboratory, the structure of an NMDA receptor complex isolated from rat brain neuronal membranes was reported to contain four protein subunits, 67, 57, 46 and 33 kDa (54). The isolated proteins apparently formed a functional NMDA receptor complex which exhibited glutamate and glycine–activated binding of [^{3}H]TCP. These observations of the biochemical nature of what might be a population of NMDA receptors are consistent with the idea of the presence of heteromeric NMDA receptors in neuronal synaptic membranes.

Further support for the hypothesis that at least a class of NMDA receptors are formed by the interaction of discrete subunits of molecular sizes between 30 and 70 kDa has come from the studies performed in our laboratories. Initially, a 71 kDa glutamate–binding protein (GBP) was isolated from synaptic plasma membranes (55). A 63 kDa protein was co-purified with the 71 kDa protein. Complete separation of the 71 from the 63 kDa protein was achieved by high performance liquid chromatography (HPLC) on TSK 3000 columns. The purified 71 kDa protein had glutamate binding activity essentially identical to the fraction containing both proteins (55). These observations together with recent evidence from studies with polyclonal antibodies raised against the 71 and 63 kDa proteins (56), have led to the conclusion that the 63 kDa protein is a deglycosylated, possibly proteolytically degraded, peptide derived from the 71 kDa protein.

Following the isolation of the 71 kDa GBP, a procedure was developed for the solubilization of intact NMDA receptor complexes. This involved the use of the zwitterionic detergent CHAPS together with NH$_4$SCN, glycerol, the non–ionic detergent polyoxyethylene–10–tridecyl ether, and six protease inhibitors (57). This procedure yielded quantitative solu-bilization of membrane proteins and of the NMDA receptor complex. An ibotenate affinity chromatography procedure and a method for selective elution of CPP–binding proteins were used to obtain a high degree of purification of a protein that binds the NMDA receptor competitive antagonists. A nearly 20,000-fold enrichment of CPP-binding entities was achieved in the purified protein fraction when compared with brain homogenates (57). This fraction is highly enriched in a glycoprotein of estimated molecular size equal to 58–60 kDa. This protein has no cross-reactivity with antibodies against the 71 kDa GBP and it does not bind either glutamate, aspartate, or NMDA. Despite the fact that the solubilized membrane protein fraction retains [^{3}H]CPP binding that is sensitive to glutamate, NMDA, aspartate and ibotenate, the 58 kDa fraction has only the aminophosphonocarboxylic acid binding sites (57). The glutamate-binding and TCP-binding sites are recovered in a fraction eluted by 1 M KCl.

When a glutamate-derived ReactiGel matrix was used in the affinity chromatography step and an NMDA-containing elution buffer was applied to elute the glutamate- and NMDA-binding protein fraction from the affinity chromatography matrix, the protein fraction isolated contained glutamate-sensitive L-[^{3}H]glutamate binding, NMDA-sensitive L-[^{3}H]glutamate binding, MK-801-sensitive [^{3}H]TCP binding, 2-AP5-sensitive [^{3}H]CPP binding, and strychnine-**insensitive** [^{3}H]glycine binding proteins (58,59). It is important to note also that this procedure yields approximately equal enrichment of all these binding sites in the fraction eluted with 5 mM NMDA. There is no evidence of either kainate-sensitive or quisqualate-sensitive [^{3}H]AMPA binding, nor of [^{3}H]kainate binding, associated with this purified protein fraction which is an indication that this preparation of proteins does not contain either the kainate or AMPA receptors. Glutamate and glycine at 1 μM concentration, as well as spermine and spermidine at 100–1000 μM kinetically activate the binding of [^{3}H]TCP to the purified proteins. Finally, all of the [^{3}H]glutamate binding sites are NMDA-sensitive.

The estimated K_D value for L-glutamate binding to the NMDA-sensitive sites is 110 nM, a value that is fairly similar to the reported K_D for L-[^{3}H]glutamate binding to synaptic membrane NMDA receptors (60) and to the K_{act} for L-glutamate enhancement of [^{3}H]TCP binding to the complex isolated by Ikin *et al.* (54). The K_I for NMDA displacement of L-[^{3}H]glutamate bound to this protein complex is 0.49 μM. This is approximately one tenth of the K_i estimated for NMDA receptors in synaptic membranes but is nearly identical to the K_{act} for NMDA enhancement of [^{3}H]TCP binding to the complex isolated by Ikin *et al.* (54). The estimated K_D for [^{3}H]TCP binding to the complex is 56 nM, a value that is comparable to that determined for binding of this ligand to synaptic membranes (43,61). The estimated K_D for [^{3}H]glycine binding to the isolated protein complex is 3.6 μM, a value that is tenfold higher than the reported K_D for the binding of this ligand to synaptic membrane receptors (62). The presence of NMDA-binding sites that have a higher affinity and of glycine-binding sites with lower affinity for the respective ligands than synaptic membrane receptors, may be an indication that the isolated proteins represent an uncommon NMDA receptor-like complex which is characterized by high affinity NMDA and low affinity glycine binding sites.

The protein fraction eluted from the L-glutamate-derivatized ReactiGel matrix by introducing 5 mM NMDA into the elution buffer contains proteins with estimated molecular weights based on migration in SDS-PAGE that are equal to 70, 62, 43 and 41 (doublet) and 36 and 31 (doublet) kDa. The proteins around 41–43 and 31–36 kDa frequently migrate on SDS-PAGE either as a diffuse band or as a doublet. This may be the result of either the presence of multiple protein species of that molecular size, or microheterogeneity of these proteins, or differential effects on protein mobility of detergents used during the purification procedure. The molecular size of the proteins in the NMDA-eluted fraction matched fairly closely the size of the complex of proteins (67, 57, 46, and 33 kDa) isolated by Ikin *et al.*, (54). L-Glutamate and glycine as well as spermine and spermidine produce an increase in the binding of [^{3}H]TCP to its recognition sites and this increase is due to both altered kinetics of ligand binding and a change in maximum binding at equilibrium. These observations are indicative of interactions between the glutamate, glycine, polyamine and TCP binding entities, i.e., the presence of a receptor complex in the purified protein fraction. When the NMDA-eluted fraction was subjected to sucrose density sedimentation in 5–20% linear sucrose gradients, to Sephacryl S–400 HR gel permeation chromatography, or to HPLC on TSK 3000 size exclusion columns, the complex of proteins remained intact without evidence of dissociation of the protein subunits. The estimated Stokes radius for this complex of proteins was close to that of a standard protein of molecular size equal to 260 kDa. Although no corrections have yet been made for the possible association of detergent molecules to this protein complex, it is likely that the molecular size of this multi–subunit structure is greater than 200 kDa.

In reconstitution studies performed using the Takagi–Montal cell with planar lipid bilayer membranes formed from liposome–reconstituted purified proteins, glutamate (8 µM L–glutamate) activates ion channels with multiple conductance states. The predominant states are of 21 pS and 45 pS conductance. Glutamate at concentrations as low as 300 nM produced detectable single or multiple ion channel responses. From the relationship between glutamate concentration and integrated current through the ion channels, an estimate of the K_{act} equal to 0.8 µM for L–glutamate was obtained. This K_{act} is very similar to the constant estimated for [^{14}C]methylamine flux into liposomes reconstituted with partially purified preparations of the glutamate–binding proteins (0.32 µM) (63) and to the estimated K_{act} for glutamate activation of NMDA receptors in intact neurons (1.1 µM) (64). NMDA as well as glutamate induced these channel opening events and the agonist–activated ion channels in these planar bilayer membranes were blocked by 2–AP5, ketamine, MK–801 and elevated concentrations of Mg^{2+} (65).

3. Cloning of NMDA Receptor Protein cDNAs

A very different approach from the direct biochemical purification and characterization of putative receptor proteins was used by Moriyoshi *et al.* (66) to clone the cDNA for a functional NMDA receptor protein. These investigators employed functional expression cloning strategies to identify a cDNA clone from a rat brain cDNA library. Frog oocytes injected with cRNA synthesized from this clone exhibited NMDA and L–glutamate–activated ion currents that were sensitive to inhibition by the competitive antagonists of NMDA receptors, the phosphonoaminocarboxylic acids, required glycine for full activation of channels by the agonists, and the fully activated channels were blocked by non–competitive antagonists such as MK–801 and Mg^{2+}. All aspects of the function of these NMDA–activated ion currents, including the apparent high conductance of the channels for Ca^{2+}, were very similar to the characteristics exhibited by NMDA receptor–ion channels in brain neurons. This cloned form of the NMDA receptor was named the NMDAR1.

The major difference between the channels formed by expressing the NMDAR1 clone and the NMDA receptor–ion channels formed in oocytes following injections of whole brain mRNA, was the lower conductance of the channels formed by the NMDAR1 clone in comparison to the conductance of the channels formed when whole brain mRNA was used for the injections (66). Shortly after this successful cloning of the NMDAR1, two research teams reported on the cloning of isoforms of the NMDA receptor, called the NMDAR2 form (67–69). Although expression of each one of the NMDAR2 clones by itself in frog oocytes or mammalian kidney cells did not lead to the formation of functional NMDA receptor–ion channels, when each of the NMDAR2 forms was co–expressed with NMDAR1 in frog oocytes or kidney cells they formed receptor complexes with ion channels that had much higher conductances (10 to 500 times higher) than those observed for channels formed by expression of the NMDAR1 cloned species alone. In addition to the markedly greater conductance of these channels detected following the co–expression of NMDAR1 and NMDAR2 forms of the receptor proteins, the receptors formed as hybrids of NMDAR1 and R2 also exhibited varying degrees of sensitivity to the co–agonist glycine and to the competitive antagonist 2–AP5 and non–competitive antagonist Mg^{2+} (68–69). Additional variation in the structure of receptors formed by the NMDAR1 and NMDAR2 family of proteins was revealed when alternatively spliced variant forms of NMDAR1 were discovered (70–72). At least 7 splice variant forms of NMDAR1 have been identified, two with distinguishable sensitivities to agonists, antagonists, and activators such as the polyamine spermine (70–71). The NMDAR2 exists also in four different forms, NMDAR2 A to D, that have unique patterns of expression in rodent brain and impart unique ion channel properties to the receptors formed by co–expressing NMDAR1 with each one of these isoforms of NMDAR2 (67–69). The cloned NMDAR1 and NMDAR2 forms of the receptor proteins and the variety of functional channels formed by the co–expression of isoforms of these proteins are a clear indication of the rich variability in NMDA receptor forms in brain. The combination of splice variants and alternate subunits within the family of NMDAR1 and NMDAR2 receptors has yielded extremes of receptor–ion channel characteristics that include among others the complete lack of sensitivity of the NMDAR1b splice variant to the polyamine spermine (71) and the nearly full activation of the receptor–ion channel formed from the combination of NMDAR1+NMDAR2C by glycine alone (68)!

Although no clear correlation exists between expression of the various forms of NMDAR1/NMDAR2 in brain and the previously reported heterogeneity of NMDA receptor ligand binding sites (45,46), it is conceivable that some of the receptor heterogeneity in mammalian brain may be due to the expression in different combinations of the variant forms of NMDAR1 and NMDAR2. It is also possible that there are other forms of NMDA receptors yet to be identified. For example, the inferred structure of the NMDAR1 and NMDAR2 proteins is that of large molecules of 100 to 160 kDa size. Yet, in most protein purification efforts these large molecular size proteins were not detected as entities that had glutamate, CPP or TCP/MK–801 ligand binding sites (54,58,59). The results from protein purification studies may be indicative of the lability of the large molecular size proteins that form NMDA receptors, i.e., these proteins are susceptible to proteolysis. Therefore, the smaller molecular weight proteins isolated by standard biochemical approaches that have the characteristics of NMDA receptors may represent degradation products of the larger proteins whose identity is inferred from cloning experiments. If this is indeed the case, then one would expect a substantial degree of homology or identity between the 30 to 70 kDa proteins and the 100 to 160 kDa NMDAR1 and NMDAR2 proteins. However, cloning of the cDNA for the small molecular size proteins has revealed that these proteins belong to a completely distinct family of putative receptor proteins (58,59,73).

Antibodies raised against the purified 71 kDa GBP were used to screen a rat hippocampal cDNA expression library and to clone a cDNA for this protein (58). This clone will be referred to as NMDARA1, for NMDA receptor–associated protein, and is 1.8 kb in size and yielded an antibody–labeled peptide of about 60 kDa on Western blots. The deduced amino acid sequence of the glutamate–binding protein is 516 residues long with an estimated molecular weight of 57,020. The cloned NMDARA1 protein was purified from *E. coli* extracts and it has an estimated K_D for L–glutamate binding equal to 263 nM, which is approximately equal to the K_D for the glutamate–binding protein purified from rat brain synaptic membranes. Northern blot analyses also indicate that the protein is expressed only in brain. Expression of the protein in brain and not in other tissues fits with the immunochemical data reported previously for this protein (56).

Sequence analysis of the inferred structure of the protein reveals that it contains four major hydrophobic domains of sufficient length to be considered potential transmembrane–spanning regions of the NMDARA1. However, there is no substantial homology between the glutamate–binding protein and the NMDAR1/NMDAR2 family of receptor proteins or the kainate/AMPA or the metabotropic glutamate receptor. A second protein of this NMDA receptor complex, the ~58 kDa CPP–binding subunit, was recently cloned (73) and sequence analysis of this protein also does not indicate any homology to the NMDAR1 and NMDAR2 receptor proteins. However, there is a reasonable level of homology (approximately 30%)

124

between the newly cloned CPP–binding subunit (NMDARA2) and the glutamate–binding subunit NMDARA1 (unpublished observations).

It is apparent, therefore, that both the glutamate–binding and the CPP–binding subunits of a putative NMDA receptor complex do not belong to the family of the NMDAR1/NMDAR2 receptors described above. This may be an indication that either mammalian brain neurons express two very distinct families of NMDA receptors or that the complex of proteins with molecular sizes in the range of 30 to 70 kDa do not normally function as NMDA receptors in neurons. Absolute evidence that these proteins can function as NMDA receptors in neurons will have to await the cloning and expression of all subunits that make up the complex that was described by Ikin and colleagues (54) and by Kumar and colleagues (58). However, there is substantial, albeit indirect, evidence that suggests that this complex does have a function as an NMDA receptor–ion channel in brain neurons.

4. Evidence that the 71 kDa Glutamate–Binding Protein is Part of an NMDA Receptor

Immunoaffinity chromatography of solubilized synaptic membrane proteins through columns to which polyclonal antibodies of the 71 kDa glutamate–binding protein were attached leads to the extraction of approximately 55 to 75% of the total glutamate binding entities associated with synaptic membranes and of the purification of the 71 and 63 kDa GBPs (56,74). Electron microscopy combined with gold particle immunohistochemistry was used to show that the sites labeled by the antibodies are entities present either on the surface of or within the post–synaptic membranes and associated densities of brain synaptosomes (56). Labeling is along dendrites and around neuronal cell bodies (56).

Expression of the 71 kDa protein determined by immunohistochemical and Western blot analyses using a monoclonal antibody follows a similar time course during development of cultured hippocampal neurons as the appearance of sensitivity to cytotoxicity induced by NMDA in these neurons (75). Pre–treatment of hippocampal cell cultures with the anti–GBP antibody blocked nearly 90% of the neurodegeneration produced by NMDA but had little effect on kainate–induced toxicity (75). Also, functional NMDA receptors appear in greater numbers in cerebellar granule cells cultured under conditions that promote cell survival (elevated K^+ to 25 mM or in the presence of 140 μM NMDA) than in cells grown under conditions which do not promote cell survival (K^+ at 10 mM in extracellular medium), and this enhanced expression of functional NMDA receptors coincides with an increase in the expression of the 71 kDa protein (76). Finally, in recent studies with the hippocampal neurons it has been observed that exposure of these neurons to basic fibroblast growth factor (bFGF) protects neurons from glutamate and NMDA toxicity selectively, and that protection correlated with the near disappearance of immunochemically and immunohistochemically–detected GBP labelling in hippocampal neurons (77). All these observations fit with the idea that the 71 kDa GBP or NMDARA1 is the glutamate recognition subunit of an active NMDA receptor complex expressed in mammalian brain neurons.

5. Summary

The data from purification, ligand binding, reconstitution and immunochemical studies indicate that there is a group of small molecular size proteins (30 to 70 kDa) that form what appear to be NMDA receptor complexes. Based on cell biological studies of the expression and localization of one of the subunits of this complex, the glutamate–binding subunit, it appears that this putative NMDA receptor plays a key role in neuronal sensitivity to NMDA and in neuronal survival in early development. However, brain neurons quite clearly express another family of proteins which have all functional characteristics of an NMDA receptor plus a great degree of variability that can account for the varieties of NMDA receptors found in brain. This family of NMDA receptors, the NMDAR1 and NMDAR2, are not homologous to the small molecular size proteins of the previously described receptor complex that was isolated from synaptic membranes. If brain neurons are indeed expressing two very diverse families of proteins that function as glutamate/NMDA receptors, this must be an indication that either there is a very selective expression of one of these forms in specific neurons or neuronal compartments, or that one of these forms of the receptor plays an important role in unique functions of the cell, such as synaptic plasticity or neurodegeneration. As more information is gathered about the structure and function of these receptors, a better understanding of the expression and role of these families of receptor proteins in normal neuronal excitability will be achieved. This enhanced level of understanding about the function and activity of these receptors will be needed in order to identify more precisely the NMDA receptors most affected in pathological conditions such as hyperammonemia.

ACKNOWLEDGMENTS. This research was supported by grants from NIAAA (AA04372), ARO (DAAL03–91–G–0167), Parke–Davis and Marion Merrell Dow Scientific Educational Partnership, and by the Center for Biomedical Research, University of Kansas. I thank Kim Bland and Nancy Harmony for their assistance in the preparation of this manuscript.

References

1. Fonnum, F., 1984, J. Neurochem. **42:**1–11.
2. Cotman, C. W., and Iversen, L. L., 1987, *Trends Neurosci.* **10:**263–265.
3. Cotman, C. W., Monaghan, D. T., Ottersen, O. P., and Storm–Mathisen, J., 1987, *Trends Neurosci.* **10:**273– 280.
4. Collinridge, G. L., and Bliss, T. V. P., 1987, *Trends Neurosci.* **10:**288–293.
5. Storm–Mathisen, J., and Ottersen, O. P., 1988, in *Neurotransmitters and Cortical Function*, M. Avoli, T.A. Readre, R. W. Dykes, and P. Gloor, eds., pp. 39–70, Plenum Press, New York.
6. Nowak, L., Bregestovski, P., Ascher, P., Herbet, A., and Prochiantz, A., 1984, *Nature* **307:** 462–465.
7. MacDermott, A. B., Mayer, M. L., Westbrook, G. L., Smith, S. J., and Barker, J. L., 1986, *Nature.* **321:**519–522.
8. Sugiyama, H., Ito, I., and Watanabe, M., 1989, *Neuron.* **3:**129–132.
9. Murphy, S. N., Thayer, S. A., and Miller, R. J., 1987, *J. Neurosci.* **7:**4145–4158.
10. Xie, X., Berger, T. W., and Barrioneuevo, G., 1992, *J. Neurochem.* **67:**1009-1–13.
11. Choi, D. W., 1987, *J. Neurosci.* **7:**369–379.
12. Garthwaite, G., and Garthwaite, J., 1986, Neurosci. Lett. **66:**193–198.
13. Faden, A. I., Demediuk, P., Panter, S. S., and Vink, R., 1989, *Science.* **244:**798–800.
14. Phillis, J. W., and Walter, G. A., 1989, *Neurosci. Lett.* **106:**147–151.
15. Choi, D. W., 1990, *J. Neurosci.* **10:**2493–2501.
16. Wieloch, T., 1985, *Science.* **230:**681–683.
17. Sloviter, R. S., 1983, *Brain Res. Bull.* **10:**675–697.
18. Marcaida, G., Felipo, V., Hermenegildo, C., Miñana, M–D., and Grisolia, S., 1992, *FEBS. Lett.* **296:**67–68.
19. Maddison, J. E., Watson, W. E. J., Dodd, P. R., and Johnston, G. A. R., 1991, *J. Neurochem.* **56:**1881– 1888.
20. Recasens, M., Guiramand, J., Nourigat, N., Sassetti, I., and Deviliers, G., 1988, *Neurochem. Internat.* **13:**463–467.
21. Cull–Candy, S. G., and Usowicz, M. M., 1987, *Nature.* **325:**525–528.

22. Jahr, C. E., and Stevens, C. F., 1987, *Nature* **325:**522–525.
23. Watkins, J. C., Krogsgaard–Larsen, P., and Honoré, T., 1990, *Trends Pharmacol. Sci.* **11:**25–33.
24. Balazs, R., Jorgensen, O. S., and Hack, N., 1988, *Neuroscience* **27:**437–451.
25. Choi, D. W., Koh, J–Y., and Peters, S., 1988, *J. Neurosci.* **8:**185–196.
26. Mattson, M. P., Lee, R. E., Adams, M. E., Guthrie, P. B., and Kater, S. B., 1988, *Neuron.* **1:**865–876.
27. Mattson, M. P., Guthrie, P. B., Hayes, B. C., and Kater, S. B., 1989, *J. Neurosci.* **9:**1223–1232.
28. Lehman, J., Schneider, J., McPherson, S., Murphy, D. E., Bernard, P., Tsai, C., Bennet, D. A., Pastor, G., Steel, D. J., Boehm, C., Cheney, D. L., Liebman, J. M., Williams, M., and Wood, P. L., 1987, *J. Pharmacol. Exp. Ther.* **240:**737–746.
29. Perkins, M. N., Collins, J. F., and Stone, T. W., 1982, *Neurosci. Lett.* **32:**65–68.
30. Davies, J., Evans, R. H., Herrling, P. L., Jones, A. W., Olverman, H. J., Pook, P., and Watkins, J. C., 1986, *Brain Res.* **382:**169–173.
31. Mayer, M. L., Vyklicky, L., Jr., and Clements, J., 1989, *Nature* **338:**425–427.
32. Snell, L. D., Morter, R. S., and Johnson, K. M., 1988, *Eur. J. Pharmacol.* **156:**105–110.
33. Johnson, J. W., and Ascher, P., 1987, *Nature.* **325:**529–531.
34. Evans, R. H., Francis, A. A., and Watkins, J. C., 1977, *Experientia Basel.* **33:**489–491.
35. Mayer, M. L., and Westbrook, G. L., 1987, *Prog. Neurobiol.* **28:**197–276.
36. Anis, N. A., Berry, S. C., Burton, N. R., and Lodge, D., 1983, *Br. J. Pharmacol.* **79:**565–575.
37. Berry, S. C., Dawkins, S. L., and Lodge, D., 1984, *Br. J. Pharmacol.* **83:**179–185.
38. Lodge, D., Aram, J. A., Church, J., Davies, S. N., Martin, D., O'Shaughnessy, C. T., and Zeman, S., 1987, in *Excitatory Amino Acid Transmission*, Alan R. Liss, New York, pp. 83–90.
39. Honey, C. R., Miljkovic, Z., and Macdonald, J. F., 1985, *Neurosci. Lett.* **61:**135–139.
40. Coan, E. J., and Collingridge, G. L., 1987, *Br. J. Pharmacol.* **91:**547–556.
41. Cole, A. E., Eccles, C. U., Aryanpur, J. J., and Fisher, R. S., 1989, *Neuropharmacol.* **28:**249–254.
42. Wong, E. H. F., Kemp, J. A., Priestley, T., Knight, A. R., Woodruff, G. N., and Iversen, L. L., 1986, *Proc. Natl. Acad. Sci. USA.* **83:**7104–7108.
43. Kloog, Y., Haring, R., and Sokolovsky, M., 1988 *Biochemistry* **27**, 843–848.
44. Idriss, M., and Albuquerque, E. X., 1985, *FEBS Lett.* **189:**150–156.
45. Monaghan, D. T., Olverman, H. J., Nguyen, L., Watkins, J. C., and Cotman, C. W., 1988, *Proc. Nat. Acad. Sci. USA.* **85:**9836–9840.
46. Monaghan, D. T., and Beaton, J. A., 1991, *Eur. J. Pharmacol.* **194:**123–125.
47. Beaton, J. A., Stemsrud, K., and Monaghan, D. T., 1992, *J. Neurochem.* **59:**754–757.
48. Murphy, D. E., Scheider, J., Boehm, C., Lehmann, J., and Williams, M., 1987, *J. Pharmacol. Exp. Ther.* 778–784.
49. Ebert, B., Wong, E. H. F., and Krogsgaard–Larsen, P., 1991, *Eur. J. Pharmacol. – Mol. Pharmacol. Section.* **208:**49–52.
50. Honoré, T., Derjer, J., Nielsen, E. Ø., Watkins, J. C., Olverman, H. J., and Nielsen, M., 1989, *Eur. J. Pharmacol. – Mol. Pharm. Section.* **172:**239–247.
51. Wong, E. H. F., and Nielsen, M., 1989, *Eur. J. Pharmacol. – Mol. Pharm. Section.* **172:**493–496.
52. Sonders, M. S., Barmettler, P., Lee, J. A., Kitahara, Y., Keana, J. F. W., and Weber, E., 1990, *J. Biol. Chem.* **265:**6776–6781.
53. Haring, R., Kloog, Y., and Sokolovsky, M., 1986, *Biochemistry.* **25:**612–620.
54. Ikin, A. F., Kloog, Y., and Sokolovsky, M., 1990, *Biochemistry.* **29:**2290–2295.
55. Chen, J. W., Cunningham, M. D., Galton, N., and Michaelis, E. K., 1988, *J. Biol. Chem.* **263:**417–427.
56. Eaton, M. J., Chen, J. W., Kumar, K. N., Cong, Y., and Michaelis, E. K., 1990, *J. Biol. Chem.* **265:**16195–16204.
57. Cunningham, M. D., and Michaelis, E. K., 1990, *J. Biol. Chem.* **265:**7768–7778.
58. Kumar, K. N., Tilakaratne, N., Johnson, P. S., Allen, A. E., and Michaelis, E. K., 1991, *Nature.* **354:**70–73.
59. Michaelis, E. K., Michaelis, M. L., Kumanr, K. N., Tilakaratne, N., Joseph, D. B., Johnson, P. D., Babcock, K. K., Schowen, R. L., Minami, H., Sugawara, M., Odashima, K., and Umezawa, Y., 1992, *N.Y. Acad. Sci.* **654:**7–18.
60. Monaghan, D. T., and Cotman, C. W., 1986, *Proc. Nat. Acad. Sci. USA*, **83:**7532–7536.

61. Largent, B. L., Gundlach, A. L., and Snyder, S. L., 1986, *J. Pharmacol. Exp. Ther.* **238:** 739–748.

62. Kessler, M., Terramani, T., Lynch, G., and Baudry, M. J., 1989, *J. Neurochem.* **52:**1319– 1328.

63. Ly, A. M., and Michaelis, E. K., 1991, *Biochemistry.* **30:**4307–4316.

64. Patneau, D. K., and Mayer, M. L., 1990, *J. Neurosci.* **10:**2385–2399.

65. Eggeman, K. T., Aistrup, G., Kumar, K. N., Michaelis, E. K., and Schowen, R. L., 1991, *Neurosci. Abs.* **17:**74.

66. Moriyoshi, K., Masu, M., Ishii, T., Shigemoto, R., Mizuno, N., and Nakanishi, S., 1991, *Nature.* **354:**31–36.

67. Meguro, H., Mori, H., Araki, K., Kushiya, E., Kutsuwada, T., Yamazaki, M., Kumanishi, T., Arakawa, M., Sakimura, K., and Mishina, M., 1992, *Nature.* **357:**70–74.

68. Katsuwada, T., Kashiwabuchi, N., Mori, H., Sakimura, K., Kushiya, E., Araki, K., M eguro, H., Masaki, H., Kumanishi, T., Arakawa, M., and Mishina, M., 1992, *Nature.* **358:**36–41.

69. Monyer, H., Sprengel, R., Schoepfer, R., Herb, A., Higuchi, M., Lomeli, H., Burnashev, N., Sakman, B., and Seeburg, P., 1992, *Science.* **256:**1217–1221.

70. Nakanishi, N., Axel, R., and Shneider, N. A., 1992, *Proc. Natl. Acad. Sci. USA.* **89:**8552– 8556.

71. Durand, G. M., Gregor, P., Zheng, X., Bennett, M. V. L., Uhl, G. R., and Zukin, R. S., 1992, *Proc. Natl. Acad. Sci. USA.* **89:**9359–9363.

72. Nakanishi, S., 1992, *Science* **258:**597–603.

73. Johnson, P., Kumar, K., Ahmad, M., Wong, D., Eggeman, K., Wu, J.-P., Bigge, C., and Michaelis, E. K., 1992, *Neurosci. Abstr.* **18:**258.

74. Wang, H., Kumar, N. K., and Michaelis, E. K., 1991, *Neuroscience,* in press.

75. Mattson, M. P., Wang, H., and Michaelis, E. K., 1991, *Brain Res.* **565:**94–108.

76. Balazs, R., Resink, A., Hack, N., Van der Valk, J. B. F., Kumar, K. N., and Michaelis, E. K., 1992, *Neurosci. Lett.* **109:**113–116.

77. Mattson, M. P., Kumar, K. N., Wang, H., Cheng, B., and Michaelis, E. K., 1993, Manuscript submitted.

Ganglioside GM1 and its Semisynthetic Lysogangliosides Reduce Glutamate Neurotoxicity by a Novel Mechanism

Erminio Costa, David Armstrong,
Alessandro Guidotti, Alexander Kharlamov,
Lech Kiedrowski, and Jarda T. Wroblewski

1. Introduction

In the brain interneuronal communications occur at synapses. Synapses are specialized junctions between the presynaptic terminals of the afferent neurons and the receiving part of the postsynaptic neurons: the pre– and postsynaptic components are separated by a space termed a synaptic gap. Nerve impulses reaching the presynaptic terminal release a transmitter which is stored in synaptic vesicles preferentially located in the nerve terminal. The transmitter is released into the synaptic gap and diffuses in this space thereby reaching specific sites that bind the transmitter with high affinity. These transmitter recognition sites are part of a more complex supramolecular structure termed transmitter receptor(s) which are located on the post and presynaptic component of the synaptic junction. There are two structurally and functionally distinct types of transmitter receptors (ionotropic and metabotropic) which differ by the mechanism attending the transduction of the transmitter signals into specific receptor responses. Both transduction mechanisms generate a metabolic event of high functional significance for the receiving neuron. When metabotropic receptors (Fig. 1) are activated by the transmitter, the catalytic activity of specific enzymes which are coupled to the receptor increases. Upon binding of the transmitter, greater amounts of the products of the receptor–linked enzyme (second messenger) are produced inside the receiving neuron. This messenger diffuses into accessible intraneuronal metabolic compartments of the receiving neuron thereby activating a specific key enzyme (often a protein kinase) which triggers a cascade of metabolic events.

The ionotropic receptor (Fig. 2) includes an ion selective channel which opens when gated by the transmitter. Ionotropic receptors are heterooligomeric protein complexes that include a number of subunits. In each subunit we distinguish extra– and intraneuronal domain which are linked by 4 helical transmembrane hydrophobic amino acid chains that form the ion pore. Often an ionotropic receptor includes five subunits. Thus, the twenty hydrophobic helices included in an ionotropic receptor together form the ion selective pore which when gated by the transmitter allows the influx or efflux of specific ions within a time

Fidia–Georgetown Institute for the Neurosciences, 3900 Reservoir Road, N.W., Washington, D.C., 20007, USA

Cirrhosis, Hyperammonemia, and Hepatic Encephalopathy,
Edited by S. Grisolia and V. Felipo, Plenum Press, New York, 1994

129

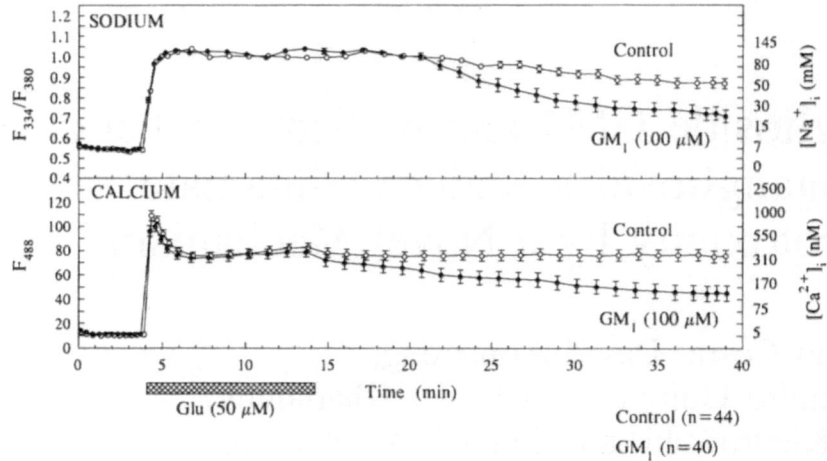

Figure 1. Signal Transduction at Metabotropic Glutamate Receptors.

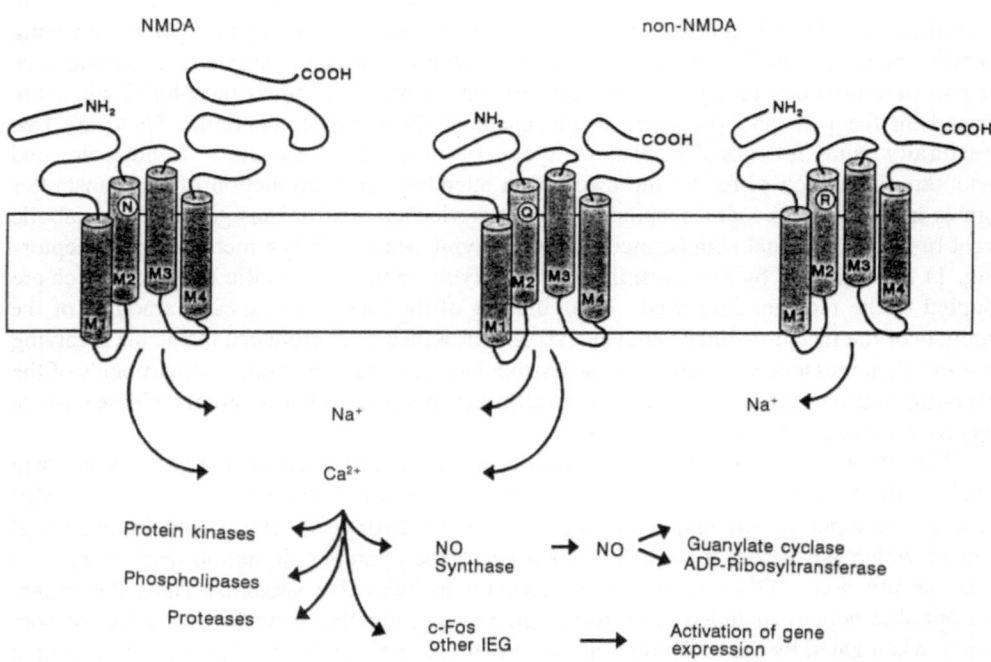

Figure 2. Signal Transduction at Ionotropic Glutamate Receptors.

constant measured in milliseconds. This ion selectivity depends on the composition of an amino acid ring, situated in the ion pore gated by the transmitter.

The transsynaptic activation of ionotropic receptors via the neurotransmitter acetylcholine or selected transmitter amino acids can mediate a rapid interneuronal communication by eliciting repetitive bursts of specific ion fluxes through the ionotropic receptors included in the postsynaptic receptor cluster. These can be monitored as electrical currents lasting from a fraction of a millisecond to several milliseconds. Two kinds of elementary events participate in this interneuronal exchange of information (1): a) the quantal release of the neurotransmitter from the vesicular stores located on one side of the synaptic gap and b) the random opening of single channels associated with the ionotropic receptors that are clustered in the neuronal membrane facing the transmitter release site or in the membrane from which the transmitter has been released (autoreceptors). Each opening of a channel is triggered by the high–affinity binding of a transmitter molecule to specific recognition sites located in the extracellular domain of the ionotropic receptor (Fig. 2). Depending on the rate of release of transmitter, the inactivation of the transmitter, the affinity constant of receptor recognition site, and on the intrinsic desensitizing capacity of the receptor, each channel can respond repetitively with a characteristic kinetic to a quantal release of transmitter. The intensity of the current thereby generated is due to a synchronomous flux of ions through the open channels. Hence, the elementary characteristic of the channel determines the rate of ion fluxes during each opening, its open time and its frequency of opening. These properties characterize the current intensity of the receptor response (2). The ionic characteristics of the current are determined by the ionic specific permeability of the channel which in turn is related to the amino acid sequence included in the hydrophobic helices located in the transmembrane receptor domain (Table 1). Inhibitory hyperpolazing currents usually are determined by either an efflux of K^+ or an influx of Cl^-, whereas, excitatory depolarizing currents include an influx of Na^+ that can be associated with a variable amount of Ca^{2+} influx. In excitatory amino acid receptors the characteristics of the amino acids located in a strategic site of the twenty hydrophobic transmembrane helices determines the Ca^{2+} and Mg^{2+} permeability of the receptor (3,4). The current profile generated eminently depends on the structure of the receptors forming the receptor cluster in the postsynaptic membrane. It appears that these clusters include a mosaic of structurally different receptors and it is possible that there is a dynamic in this receptor cluster heterogeneity (1) which might be regulated functionally.

2. Ca^{2+} as a Second Messenger in Synaptic Signal Transduction

Ionotropic receptors are classified as cationic or anionic depending on their specific ion permeability characteristics. The receptors for the amino acid transmitter GABA (γ-aminobutyric acid) include anionic (Cl^-) channels, whereas, those for the excitatory neurotransmitter glutamate include cationic channel permeable to Na^+ and depending on the structure of the transmembrane domain they also are permeable to Ca^{2+} (4).

If Ca^{2+} permeable cationic channels contribute to the mosaic of glutamate ionotropic receptor clusters, the amount of the Ca^{2+} entering the neuron depends on the number and kinetic properties of the Ca^{2+} permeable channels. There is a great structural variability of glutamate ionotropic receptors (Table 1). Some glutamate ionotropic receptors have a high conductance (60pS) and are termed NMDA selective glutamate receptors for their ability to be gated by the synthetic agonist N–methyl–D–aspartate. Others have a low conductance (10–20 pS) and are termed from their capability to be selectively activated by the synthetic agonists kainic acid, or AMPA (α–amino–3–hydroxy–5–methyl–4–isoxazole) (Table 1). The AMPA selective glutamate activated channels similarly to the voltage–dependent Ca^{2+} channels desensitize rapidly. In contrast, the NMDA selective glutamate channels are less resilient to desensitization (5).

TABLE 1. Gene diversity for glutamate ionotropic receptor subunits.

Subunit Gene	KDa Approximate	Homology to GLU R1 %	Specific Agonist	Cation Selectivity of Recombinant Homomeric Glutamate Receptors
GLU R1 GLU R2 GLU R3 GLU R4	100	68–73	AMPA	Ca^{2+}, Mg^2 Na^+ Ca^{2+} Ca^{2+}
GLU R5 GLU R6 GLU R7	100	38–40	Kainate	None Na^+ Ca^{2+} Mg^{2+} None
KA 1 KA 2	100	35	Kainate	None
NMDA R1	160	22		Ca^{2+}
NMDA R2A NMDA R2B NMDA R2C	160 160 133	12–18	NMDA	None

The variability of natural receptor is further complicated by various posttranscriptional mechanisms (Flip Flop, mRNA, editing etc...) (see Sommer et al., reference #4).

Since in the cytoplasm of neurons the free ionized Ca^{2+} content ($[Ca^{2+}]i$) is about 50 nM even a small influx of Ca^{2+} through a cluster of Ca^{2+} permeable channels may increase the $[Ca^{2+}]i$ concentration of a given intraneuronal compartment, significantly (6). This increase acts as a second messenger by activating various Ca^{2+} dependent enzymes that are present in that compartment (6). The activation of various molecular forms of Ca^{2+} dependent protein kinases (Ca^{2+}–PK) often leads to their translocation to neuronal membranes (7) or to the nucleus. In their novel neuronal membrane milieu these Ca^{2+}–PKs modulate directly or indirectly the catalytic activity of Na^+/K^+–ATPases, Ca^{2+}–ATPase and the Na^+/Ca^{2+} exchanger, all of which very likely contribute to the Ca^{2+} extrusion efficacy and thereby regulate the $[Ca^{2+}]i$ homeostasis. Moreover, in the endoplasmic reticulum of neurons and glial cells there are important Ca^{2+} stores regulated by a specific Ca^{2+} channel modulated by inositol triphosphate (IP_3) (8). Some of these Ca^{2+} stores are also released by increases of $[Ca^{2+}]i$, hence, the Ca^{2+} influx from extracellular fluid via ionotropic Ca^{2+} channels can further increase the $[Ca^{2+}]i$ levels by releasing Ca^{2+} from endoplasmic reticulum stores (8).
Hence the wave of $[Ca^{2+}]i$ increase that is generated by the activation of Ca^{2+} permeable glutamate receptors, diffuses through the neuroplasm at times reaching even the nucleus. The ratio between glutamate mediated Ca^{2+} influx and efficiency of homeostatic regulation will determine the extent and diffusion of the neuroplasmic $[Ca^{2+}]i$ wave elicited by glutamate receptor stimulation.

3. Third Nuclear Messengers in the Sequelae of Synaptic Signal Transduction

Ca^{2+}–PKs when activated by an appropriate accumulation of $[Ca^{2+}]i$ can translocate to the nucleus and activate the transcription of immediate early genes (IEG) which encode for proteins that revert to the nucleus and activate the synchronous transcription of a number of genes including in their 5'upstream regulatory region specific DNA motifs that recognize homo− or heterodimers of IEG encoded proteins (9). Some of these proteins such as the cyclic AMP responsive element (CRE) activate transcription only after they are phosphorylated by the cyclic AMP−dependent protein kinase. Different genes include in their 5'upstream regulatory domain several DNA motifs with a specific high affinity for various IEG encoded proteins. The binding of specific heterodimeric complexes of these proteins expresses specific instructions (time and extent) to the gene regulatory region. Thus, in this information is encoded a dimension of time in terms of transcription stimulation and of intermittent pauses of transcription activation (10). Hence, IEG responses harmoniously regulate transcriptional activation of multigene programs. Reverting to the changes in the catalytic activity of Ca^{2+}−PKs and to their translocation to a new intracellular milieu caused by the rapid oscillation of $[Ca^{2+}]i$ we like to stress the functional importance of this "per se" electrically silent neuronal response which by orchestrating important changes in gene expression, regulates functionally related dynamic changes of neuronal structure (11). It is currently believed that an important goal for the transmitter mediated changes in Cl^-, K^+, and Na^+ fluxes across the neuronal membrane is the regulation of second messenger responses triggered by Ca^{2+}, including the activation or the inhibition of Ca^{2+} dependent specific transcriptional programs which are synchronized and regulated via the transcription activation of IEG cascades.

4. Polymorphism of Glutamate Ionotropic Receptors

The transmitters that function in the regulation of transmembrane rapid ion fluxes include acetylcholine, acting on brain nicotinic receptors, as well as glutamate and γ−aminobutyric acid (GABA) acting on a great variety of structurally different ionotropic receptors. The glutamate, GABA and acetylcholine receptors participating in these ionotropic signal transductions are characterized by a dynamic structural polymorphism supported by an amazing number of structurally heterogenous genes each encoding for various members of several receptor subunit families (Table 1). In turn these subunits, with their variable assembly form a great number of structurally different heterooligomeric (pentameric) ionotropic receptors. Lines of indirect experimentation suggest that some clusters of ionotropic receptors may regulate dynamically this structural heterogeneity by heterologous receptor modulation (12). Such structural variability is probably regulated via receptor−receptor interactions not only during ontogenetic neuronal differentiation but also during the normal function in the mature brain (12). In support of such functionally regulated structural variability of the clusters of amino acid transmitter gated ionotropic receptors is the evidence that in neuronal cultures of glutamatergic neurons the activation of glutamate receptors changes the content of specific mRNAs encoding for $GABA_A$ receptors (12). In considering that more than a dozen genes encode for structurally different subunits of glutamate ionotropic receptors (Table 1) it is reasonable to think that this extraordinary variability of receptor subtypes is used functionally to encode the precise tuning of the dynamic state of the ionotropic receptor mosaic expressed in subsynaptic receptor clusters located in these cells. A similar heterogeneity has been extensively documented for $GABA_A$ (13) receptors. The transmembrane domain of glutamate and other ionotropic receptor subunits usually possesses a greater intra−family homology than that of the extra− and intraneuronal domains. Moreover, for any given receptor subunit, such structural heterogeneity encoded in the multiple specific genes may be increased by alternative splicing and/or by a single amino acid substitution by mRNA editing (2,4). Such a mechanism is operative at the ionic filtering ring of AMPA and NMDA selective glutamate receptors (3).

A few years back, glutamate receptors were empirically differentiated according to their selective suceptiblity to the activation by synthetic agonists: NMDA, kainate or AMPA. We now know that to this provisional receptor classification corresponds different structural receptor characteristics (Table 1). In fact, three subunit families can be distinguished on the basis of their selective susceptibility to be activated by the three agonists. Moreover, most of the recombinant homomeric receptors formed by the members of each of the three subunit families are also selectively activated by the synthetic ligands specific for that family (Table 1). Since, we do not know how the subunits are assembled to form natural heterooligomeric receptors nor do we know the subunit stoichiometry of native receptors, the mechanisms regulating the dynamic changes occurring in postsynaptic receptor clusters are still a matter of speculation. However, since many homomeric receptors are not functional (Table 1) we can at least infer that probably native receptors are heterooligomeric structures. Moreover, the number of different genes encoding for the receptor subunits allows us to infer that a high degree of structural variability may be included in native ionotropic receptors.

5. Glutamate Metabotropic Receptors

Ten years ago the dominant assumption was that glutamate receptors were only ionotropic, even though at that time we knew that GABA receptors included two classes of signal transduction mechanisms: ionotropic ($GABA_A$) and metabotropic ($GABA_B$).

In 1986 we reported (14,15) that in brain there are expressed glutamate metabotropic receptors with signal transduction operated via G proteins linked to phospholipase C which release IP_3 from membrane inositide phospholipids following the binding of glutamate to specific receptor recognition sites.

The existence of such metabotropic receptors was confirmed by others (11,16) but a wide acceptance of this novel finding was not prompt. Recently, five structurally different metabotropic receptors activated by glutamate have been cloned (17,18). This evidence prompted acceptance of the fact that IP_3 and cyclic AMP biosynthesis participation on the transduction mechanisms of brain glutamatergic transmission (11). It appears that the widely accepted trophic actions of glutamate in CNS may be mediated by such metabotropic receptors (19). For instance, the role of metabotropic receptors in the regulation of BDNF mRNA expression in response to brain injuries has been documented, recently (20). Hence, in thromboembolic, hemorrhagic, traumatic and perhaps hepatotoxic brain damage, the associated increase of glutamate content in the interstitial fluids of the areas surrounding the primary brain injury (area penumbra) on one hand causes excitotoxicity but one the other also promotes the biosynthesis of neurotrophines that by stimulating axonal regeneration and sprouting prompt synaptic plasticity.

Recent work has begun to focus on the notion that in the area penumbra two distinct functional parts may be distinguished: one more proximal to the primary injury which is characterized by excitotoxic glutamate actions and another more distal from the injury where neurotrophic actions are operative. The latter area is further characterized by a glutamate mediated IEG transcriptional activation leading to the formation of third nuclear messengers (9). These appear to be essential in coordinating the synchronized transcription of neuronal proteins mediating genetic programs fostering the functional adaptation and compensation to the deficit elicited by the area of primary brain insult and by the successive excitotoxicity typical of the area penumbra proximal to the injury.

Since the activation of glutamate ionotropic receptor mediates IEG induction (10) it appears mandatory that antagonists of glutamate ionotropic receptors should not be used in the long-term treatment of brain injuries, in order to allow the activation of physiological compensatory mechanisms.

6. Glutamate Excitotoxicity in CNS Pathology

Glutamate is the only known transmitter that causes neurotoxicity via a persistent stimulation of Ca^{2+} permeable specific ionotropic receptors resilient to destabilization. It is widely accepted that a primary mechanism of this glutamate excitotoxicity is an increase of $[Ca^{2+}]i$, due to destabilization of mechanisms operative in $[Ca^{2+}]i$ homeostasis (21,22). Probably, the activation and membrane translocation of Ca^{2+}–PK play a role in such destabilization (23). Perhaps the phosphorylation of proteins attending to the regulation of the Ca^{2+} extrusion mechanisms (i.e., Na^+/K^+–ATPase, Ca^{2+}–ATPase, Na^+/Ca^{2+} exchanger) are impaired in their function when phosphorylated. Following acute neuronal degeneration as a result of anoxia secondary to cardiovascular failure or due to insufficiency of glutamate reuptake secondary to hepatotoxicity, the levels of glutamate in brain interstitial fluids increase causing a persistent depolarization of glutamatergic neurons thus furthering the increase of interstitial fluid glutamate content. Hence, brain areas surrounding the primary brain lesion as well as extending away from the lesion, become at risk of impending glutamate–mediated neuronal death.

As discussed earlier the use of specific glutamate receptor blockers cannot become the therapeutic approach of choice for the following reasons: a) the polymorphism of glutamate receptors (Table 1) imposes the use of a "polypharmacy" to inhibit the structurally different glutamate receptors, located in various brain structures b) glutamate via metabotropic and also ionotropic receptors causes neurotrophic actions (19,20) probably via IEG induction, thereby facilitating synaptic plasticity and functional compensation to the deficit induced by the primary lesion; c) glutamate is the transmitter involved in the signal transduction of about 30% of mammalian brain synapses, and therefore the blockade of glutamate receptors may disrupt the regulation of central cardiovascular and respiratory control mechanisms. Hence the preferred therapeutic approach should not be directed towards the block of the glutamate recognition site or of the ionotropic channel but rather should be targeted towards facilitation of $[Ca^{2+}]i$ homeostasis by activating those mechanisms that increase Ca^{2+} extrusion from neurons. Since by definition such a therapeutic approach is selectively directed to those brain areas that are affected by the excitotoxic action of glutamate, the drugs that facilitate Ca^{2+} extrusion will spare the glutamatergic transmission in the intact brain areas and thereby foster physiological compensatory mechanisms.

A similar therapeutic strategy should also be followed in treating glutamate excitotoxicity associated with epilepsy. In this epilepsy the time interval between successive depolarizations of glutamatergic neurons is shortened, and the rate of Ca^{2+} influx is increased thereby challenging the Ca^{2+} homeostatic mechanisms. This causes an accumulation of $[Ca^{2+}]i$ content in the postsynaptic neuron (24). Such an accumulation with time leads to unavoidable neuronal death. Thus, in epileptic patients convulsive discharges increase the rate of glutamate–induced neuronal death with the consequence progressive impairment of brain functions.

7. Ganglioside Inhibition of the $[Ca^{2+}]i$ Amplification During Glutamate Excitotoxicity

Connor et al. (24) reported that sphingosine inhibits the increase of $[Ca^{2+}]i$ amplification elicited by persistent activation of ionotropic glutamate receptors without inhibiting the influx of Ca^{2+}. This finding suggested to us that sphingosine reduces the Ca^{2+} accumulation without blocking the glutamate gating of Ca^{2+} channels. Thus, it appeared possible that the destabilization of $[Ca^{2+}]i$ homeostasis during excitotoxicity could be attenuated by sphingosine analogues devoid of the sphingosine primary toxic action on cell membranes. Gangliosides include in their chemical structure a sphingosine molecule but they, unlike sphingosine, are not toxic and do not alter the membrane structure. Moreover, in previous studies (7) we found that the ganglioside GM1 inhibited PKC translocation to the neuronal membrane elicited by glutamate. Collectively, these data prompted us to study whether GM1 could reduce glutamate excitotoxicity in primary neuronal cultures. The results obtained following our "in vitro" investigations clearly indicated that glutamate excitotoxicity could be inhibited by GM1 (25,26). This inhibition was dose dependent, required the presence of sialic acid and correlated with the amount of GM1 inserted in the neuronal membrane (25). Since GM1 failed to change the kinetics of the glutamate gating of NMDA selective glutamate receptors (25,26) we concluded that this drug was acting by a novel mechanism. In subsequent studies, gangliosides were found to inhibit the amplification of the $[Ca^{2+}]i$ responses by an action targeted on the mechanism of $[Ca^{2+}]i$ homeostasis (27). Presumably GM1, by inhibiting PKC translocation to the membranes, was increasing the efficacy of Ca^{2+} extrusion (23,28). This novel therapeutic strategy was termed Receptor Abuse Dependent Antagonism (RADA) to stress the selective drug action on pathological paroxysmal mechanisms of glutamate receptor activation leading to neurotoxicity (26).

The physiological importance of gangliosides in the regulation of neuronal function was inferred from the changes in the profile of gangliosides neuronal content associated with CNS development and maturation. Moreover, hereditary abnormalities in the amounts and composition of membrane gangliosides can lead to serious neuropathologies (29,30). Finally newborn infants who died of hypoxia have a decrease in brain ganglioside content that correlates with the severity of the brain damage (31). It is important to keep in mind that gangliosides added to primary neuronal cultures slowly incorporate into the neuronal membrane and participate to the metabolism and turnover of the endogenous membrane gangliosides (32,33).

As mentioned above exogenously applied gangliosides fail to change the resting characteristics of basal electrogenic membrane properties or the ion flux through cationic channels during the stimulation of glutamate receptor (26). This lack of action can be appreciated from the measurement of $[Ca^{2+}]i$ obtained by monitoring free cytosolic Ca^{2+} content by fluo 3 imaging in cerebellar neuronal cultures during persistent activation of glutamate receptors (Fig. 3). In fact, the initial Ca^{2+} fluxes elicited by excitotoxic doses of glutamate are not changed by GM1 though this ganglioside attenuates the Ca^{2+} homeostasis destabilization and protects the neuron from glutamate excitotoxicity (27).

8. Gangliosides and Lysogangliosides Bioavailability

When gangliosides are used as drugs one must consider that the physiochemical properties of gangliosides prevent an efficient bioavailability of these drugs (32). Specifically, the micellar aggregation of gangliosides in solution, limits the concentrations of gangliosides in monomeric configuration which are suitable for membrane insertion. Thus, the micellar aggregation not only slows the ganglioside insertion rate into neuronal membranes, but also impairs its absorption when given orally (Table 2). The propensity of gangliosides to form micellar aggregates prompted the synthesis of lysoganglioside derivatives which evince a higher proportion of monomeric configurations when they are in solution and thereby assure an appropriate oral absorption (Table 2). Moreover, their prevailing monomeric configuration in body fluids is greater than that of gangliosides and thereby lysogangliosides reach the brain in concentrations greater than that of natural gangliosides. Two semisynthetic derivatives of GM1 in which the fatty acid of the ceramide was eliminated leaving only the lipidic N–acetyl sphingosine moiety were found to be most effective in protecting from glutamate excitotoxicity as demonstrated following "in vitro" and "in vivo" experiments. These compounds are termed LIGA 4 (11^3Nen 5 Ac Gg Ose$_4$–2d–erythro–1,3–dihydroxy–2–

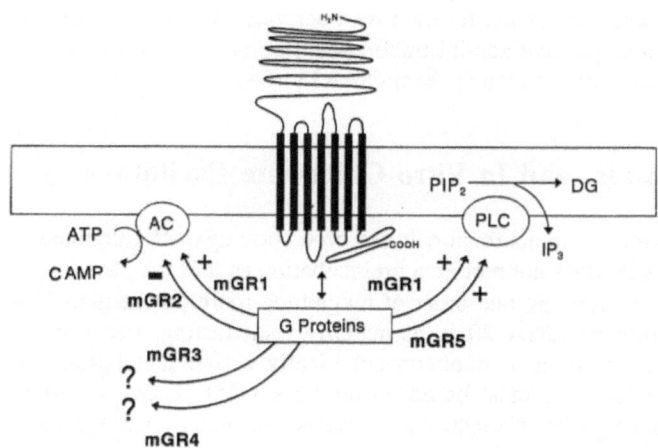

Fig 3. Effect of GM1 pretreatment on sodium and calcium extrusion after toxic glutamate exposure. Cerebellar granule (7 days in culture) were treated with 100 μM GM_1 for 2 h, then loaded with SBFI and Fluo-3 and challenged with glutamate. Note that GM_1-pretreatment facilitates both sodium and calcium extrusion. However, GM_1 facilitation of calcium extrusion precedes by 5 min the onset of sodium extrusion facilitation. This indicates that GM1 pretreatment might enhance Ca^{2+} extrusion, possibly affecting by Ca^{2+}-ATPase and decreasing the inhibition of Na^+/K^+-ATPase. As a result sodium is extruded faster. This creates better conditions for Na^+/Ca^{2+} exchanger to extrude calcium.

Table 2. Brain and plasma ganglioside content in rats receiving 70 μmol/kg (p.o.) of each compound

Drug	Ganglioside Content (μM)	
	Brain	Plasma
GM1	0.07	0.05
LIGA 4	0.55	0.69
LIGA 20	0.86	1.30

Rats fastened for 18 hr. received ^{3}H–GM1, ^{3}H–LIGA 4, ^{3}H LIGA 20 solution (2ml) by oral gavage. Brain and plasma gangliosides (GM1, LIGA 4, and LIGA 20) content was determined after extraction and thin layer chromatography separation of the authentic drug from their metabolites with techniques similar to those described by Ghidoni et al., (32).

acetylamide–4–trans–octadecene) or LIGA 20, a N–di–chloroacetyl derivative of LIGA 4 (34).

The structural differences between LIGA 4 and LIGA 20 are manifested in different aggregation properties in aqueous solutions (33), which are more favorable to membrane insertion. In neuronal cultures, the protection against excitotoxicity by LIGA 4 and 20 was obtained with doses that are one order of magnitude lower than those of GM1 (34).

oreover, the occurrence of neuroprotective action was instantaneous as opposed to the one hour of pretreatment required to obtain maximal protection with GM1 (34). In addition, in the case of LIGA 20 the "in vitro" protection persisted longer than 24 hours. Also, "in vivo" LIGA 20 has a $t_{1/2}$ of about 36 hours which is significantly longer lasting than that of GM1. Table 2 reports the plasma ganglioside or lysoganglioside content in rats receiving GM1,

LIGA 4 and LIGA 20 (70 µmol/kg p.o.) six hours earlier. It is evident that the drug concentrations in brain are greater for the LIGA derivatives than for GM1. Moreover, in the case of LIGA 20 a single oral administration is sufficient to obtain levels (Table 2) that "in vitro" are able to attenuate glutamate excitotoxicity (34).

9. Lysogangliosides and In Vitro Glutamate Excitotoxicity

As stated previously, in neuronal cultures protection against glutamate excitotoxicity by LIGA 4 and LIGA 20 does not require a preincubation as it is the case for GM1. Moreover, the two LIGA derivatives are one order of magnitude more potent than GM1. In addition, the protective action of LIGA 20 is particularly long lasting, because this compound is metabolized by the neurons into pharmacologically active metabolites operative against glutamate neurotoxicity. It must be added that like GM1, LIGA 4 and LIGA 20 do not modify the glutamate gating of ionotropic channels. Recently a Ca^{2+}–dependent ganglioside binding protein (gangliomodulin) was isolated from a soluble fraction of mouse brain and was identified as calmodulin (35). Thus, gangliosides may regulate calmodulin dependent enzymes when they are inserted in neuronal membranes. Since Ca^{2+}–ATPase was shown to be activated by calmodulin in a Ca^{2+}–independent manner, the binding to calmodulin of gangliosides inserted in neuronal membranes may anchor calmodulin to the membrane and facilitate its catalytic activity on the Ca^{2+}–ATPase. This calmodulin–ganglioside interaction may contribute to the ganglioside antagonism against the $[Ca^{2+}]i$ homeostasis destabilization induced by glutamate.

10. In Vivo Activity of Liga Derivatives Against Photochemically Induced Thrombotic Lesion of Rat Cerebral Cortex

Several experimental models of focal brain ischemias are currently available in various animal species (36). In rat the model of cortical photothrombosis is convenient to evaluate drug efficacy in attenuating the progress of excitotoxicity in the area penumbra surrounding the primary brain infarct (37). In this model the skull of the rat is exposed under general anesthesia before the injection of the Rose Bengal dye (80 mg/kg i.v.) and subsequently the skull is irradiated with a beam of cold white light. The light source consisted of the power supply (single output rated 15 VDC: 8.4 amp to 10.4 amp), a fan and a dichroic halogen bulb with parabolic reflector (12 V 100 w, wavelength 400–1200 nM with peak energy at 1000 nM, 3400 degrees Kelvin). The interaction of the light with the dye releases oxygen singlets that mediate the endothelial cell damage leading to photothrombosis. The end result is a reproducible focal cortical infarct. A light beam of 3 mm was centered 1.8 mm posterior to the Bregma and 2.8 mm lateral to the midline, and corresponds to a location overlying the parietal sensory motor cortex. Following light exposure ranging from 10 minutes to 1 hour, a lesion is produced in the cortex which in many respects reproduces the hemodynamic consequences reflecting those of spontaneous cortical cerebral infarction in humans (37). Vasogenic edema is detected in areas distal to the infarct which may even induce secondary ischemic brain damage. In our model, examination of Nissl stained tissue sections revealed neuronal damage and degeneration in the area penumbra (38). The severity of this lesion decreases in a manner related to distance from the infarct. The role of glutamate in the dynamic progression of area penumbra was demonstrated by the ability of two glutamateionotropic receptor antagonists (MK–801 or NBQX) to reduce the size and severity of this neuronal degeneration (38). Moreover, LIGA 4 and LIGA 20 which in contrast to MK–801 fail to inhibit glutamate channel gating also can reduce the severity of neuronal cell death in the area penumbra (Table 3). Both compounds like MK–801 reduce the translocation of PKC in area penumbra as documented by reductions in [³H] PDBU binding (Table 3). These results support the view that LIGA 4 and LIGA 20 reduce glutamate excitotoxicity by decreasing PKC activation and translocation without changing glutamate gated channel gating. Thus, they selectively act within the area penumbra and unlike MK–801 they do not impair IEG induction (cfos) which requires the function of glutamate channels to occur (Table 3). Similar protection was also obtained with pretreatment with GM1.

Table 3. Effect of MK–801, LIGA4, LIGA20 on [³H]PDBu binding, Fos expression and cell loss in the area penumbra following focal brain ischemia

	[3H]PDBu BINDING (a) (% of control)	FOS (b) EXPRESION	NUMBERS OF NEURONS (c) (% of control)
VEHICLE (control)	100#	Δ	100#
MK–801	112#	▲	120#
LIGA4	117#	Δ	112#
LIGA20	103#	Δ	104#
PHOTOCHEMICAL LESION	362*	▲ ▲ ▲	40*
LESION + MK–801	143#	▲	84#
LESION + LIGA4	207*#	▲ ▲ ▲	88#
LESION + LIGA20	241*#	▲ ▲ ▲	80#

(a) [³H] phorbol 12,13 dibutyrate binding was evaluated as optical density in the area penumbra
(b) FOS immunostaining: Δ basal level ▲ minimun increase ▲ ▲ ▲ maximun increase
(c) Nissl–positive neurons were counted in 0.125 x 0.300 mm segments of the area penumbra
# p <0.05 when compared with lesioned non treated group
* p <0.05 when compared with unlesioned vehicle treated group

11. Conclusions

Natural gangliosides (GM1) and LIGA compounds in doses that reduce the increase of [³H] PDBu binding in the area penumbra reduce also neuronal death. Unlike MK–801, they do not impair IEG induction suggesting a lack of action on physiological compensatory mechanisms deriving from the activation of specific genetic programs via IEG induction.

Thus GM1, LIGA 4 and LIGA 20 are selective antagonists of glutamate induced neuronal death, but do not alter the neurotrophic and other compensatory mechanisms triggered by glutamate. These drugs are the prototypes of a new class of therapeutic agents which reduce brain damage induced by the increase in levels of glutamate in the interstitial fluid without blocking glutamatergic transmission in either area penumbra or in the unaffected brain tissue. To stress this selective action these drugs are termed RADA (receptor abuse dependent antagonism) which defines the novel mechanism whereby they prevent glutamate excitotoxicity.

REFERENCES

1. Sakmann, B., 1992, Elementary steps in synaptic transmission revealed by currents through single ion channels, Neurons. **8**:613–629.

2. Sommer, B., Keinanen, K., Verdon, T. A., Wisden, W., Burnashev, N.,Herb, A., Kohler, M., Tagaki, T.,Sakmann, B., and Seeburg, P. H., 1990, Flip and Flop: A cell specific functional switch in glutamate operated channels of CNS. Science. **249**:1580–1585.

3. Sommer, B., Kohler, M., Sprengle, R., and Seeburg, P. H., 1991, RNA editing in brain controls a determinant of ion flow in glutamate gated channel, Cell. **67**:11–20.

4. Sommer, B., and Seeburg, P. H., 1992, Glutamate receptor channels: novel properties and new clones, TIPS **13**:291–296.

5. Hollman, M., O'Shea–Greenfield, A., Rogers, W., and Heineman, S., 1991, Ca^{2+} permeability of KA–AMPA– gated glutamate receptor channels depend on subunit composition, Science. **252**:851–853.

6. Connor, J. A., 1991, In pursuit of neuronal calcium: confessions of a bio–optician, Fidia Research Foundation Neuroscience Award Lectures, New York:Raven Press **6**:111–139.

7. Vaccarino, F., Guidotti, A., and Costa, E., 1988, Gangliosid inhibition of glutamate mediated protein kinase C translocation in primary cultures of cerebellar neurons, Proc. Natl. Acad. Sci. USA **84**:8707–8710.

8. Berridge, M. J., 1991, Phosphoinosities and cell signaling. Fidia Research Foundation Neuroscience Award Lectures. New York:Raven Press **6**:5–45.

9. Curran, T., Fos and Jun: inducible oncogenic transcription factors that function in neuronal signal transduction. (in press) Fidia Research Foundation Neuroscience Award Lectures. New York:Raven Press.

10. Szekely, A.M., Costa, E., and Grayson, D. R., 1990, Transcriptional program coordination by NMDA– sensitive glutamate receptor stimulation in primary cultures of cerebellar neurons, Mol. Pharmacol. **38**:624–633.

11. Conn, P. J., and Desai, M. A., 1991, Pharmacology and physiology of metabotropic glutamate receptorsin mammalian central nervous system, Drug Dev. Res. **24**:207–229.

12. Memo, M., Bovolin, P., Costa, E., Grayson, D. R., 1991, Regulation of γ–aminobutyric acid A receptor subunit expression by activation of N–methyl–D–aspartate–selective glutamate receptors, Mol. Pharmacol. **39**:599–603.

13. Seeburg, P. M., Wisden, W., Werden, T. A., Pritchet, D. B., Werner, P., Herb, A., Luddens, H., Sprengel, R., and Sanman, B., 1990, The GABA receptor family: Molecular and functional diversity. In: Cold Spring Harbor Symposium On Quantitative Biology. vol. LV: Cold Spring Harbor Laboratory Press pp. 29–40.

14. Nicolletti, F., Iadarola, M. J., Wroblewski, J. T., and Costa, E., 1986, Excitatory amino acid recognition sites coupled and with inosital phospholipid metabolism: Developmental changes and interaction with α_1 adreneceptors, Proc. Natl. Acad. Sci. USA **83**:1931–1935.

15. Nicoletti, F., Wroblewksi, J. T., Novelli, A., Alho, H., Guidotti, A., and Costa, E., 1986, The activation of isonitol phospholipid metabolism as a signal–transducing system for excitatory amino acids in primary cultures of cerebellar granule cells, J. Neurosci. **6**:1905–1911.

16. Sladeczek, F., Recasens, M., and Bockaert, J., 1988, A new mechanism for glutamate reception action: Phosphoinositide hydrolysis, Trends in Neurosci. **11**:545–549.

17. Masu, M., Tanabe, Y., Tsuchida, K., Shigemoto, R., and Nakanishi, S., 1991, Sequence and expression of a metabotropic glutamate receptor, Nature **349**:760–765.

18. Tanabe, Y., Masu, M., Ishis, T., Shigemoto, R., and Nakanishi, S., 1992, A family of metabotropic glutamate receptors, Neuron **8**:169–179.

19. Siliprandi, R., Lipartiti, M., Fadda, E., Santter, J., and Manev, H., 1992, Activation of the glutamate metabotropic receptor protects retina against N–methyl–D–aspartate–toxicity, Europ. J. Pharmacol. **19**:173–179.

20. Comelli, M. C., Seren, M. S., Gudolin, D., Manev, R. M., Favaron, M., Rimland, J. M., Canella, R., Negro, A., and Manev, H., 1992, Photochemical stroke and brain–derived neurotrophic factor (BDNF) mRNA expression, Neuroreport **3**:437–476.

21. Choi, D. W., 1988, Ionic dependence of glutamate neurotoxicity, J. Neurosci. **7**:369–379.

22. Choi, D. W., 1991, Excitatory amino acid neurotransmitters: Anatomical systems. In B.S. Meldrum

(Ed.), Excitatory Amino Acid Antagonists, Blackwell Scientific Publications: Oxford, pp. 14–38.

23. Favaron, M., Manev, H., Siman, R., Bertolino, M., Szekely, A. M., de Erausquin, G., Guidotti, A., and Costa, E., 1990, Down regulation of protein kinase C protects cerebellar granule neurons in primary culture from glutamate induced neuronal death, Proc. Natl. Acad. Sci. USA **87:**1983–1987.

24. Connor, J. A., Wadman, W. J., and Hockberger, P. E., 1988, Sustained dendritic gradients of Ca^{2+} induced by excitatory amino acid in CAI hippocampal neurons, Science **240:**649–653.

25. Favaron, M., Manev, H., Alho, H., Bertolino, M., Ferret, B., Guidotti, A., and Costa, E., 1988, Gangliosides prevent glutamate and kainate neurotoxicity in primary neuronal cultures of neonatal rat cerebellum and cortex, Proc. Natl. Acad. Sci. USA **85:**7351–7355.

26. Manev, H., Costa, E., Wroblewski, J. T., and Gudiotti, A., 1990b, Abusive stimulation of excitatory amino acid receptors: A strategy to limit neurotoxicity, J. Faseb. **4:**2789–2797.

27. De Erausquin, G., Manev, H., Guidotti, A., Costa, E., and Brooker, G., 1990, Gangliosides normalize distorted single cell intracellular free Ca^{2+} dynamics after toxic doses of glutamate in cerebellar granule cells, Proc. Natl. Acad. Sci. USA **87:**8017–8021.

28. Candeo, P., Favaron, M., Lengyel, I., Manev, R. M., Rimland, J. M., and Manev, H., (in press) Pathological phosphorylation causes neuronal death: Effect of okadaic acid in primary culture of cerebellar granule cells, J. Neurochem.

29. Baumann, N., Harpin, M. L., Jacque, C., 1980, Brain gangliosides in shiverer mouse: Comparison with other dysmyelinated mutants, quaking and jumpy. In: Neurological Mutations Affecting Myelination, N. Baumann (Ed.), INSERM Symposium 14, Elsevier North Holland: Amsterdam, pp. 257–262.

30. Purpura, D., 1978, Ectopic dendritic growth in mature pyromidal neruons in human ganglioside storage disease, Nature **276:**520–521.

31. Qi, Y., and Xue, Q. M., 1991, Ganglioside levels in hypoic brains from neonatal and premature infants, Mol. and Chem. Neuropathology **16:**87–95.

32. Ghidoni, R., Riboni, L., and Tettamanti, G., 1989, Metabolisms of exogenous gangliosides in cerebellar granular cells differentiated in culture. J. Neurochem. **53:**1567–1574.

33. Sonnino, S., Cantu, L., Corti, M., Acquotti, D., Kirschner, G., and Tettamanti, G., 1990, Aggregation properties of semisynthetic GM_1 ganglioside (11^3 Neu 5-Ac Gg Ose_4 Cer) containing an acetyl group as acylmoiety, Chem. Phys. Lipids **56:**49–57.

34. Manev, H., Favaron, M., Vicini, S., Guidotti, A., and Costa, E., 1990, Glutamate induced neuronal death in primary cultures of cerebellar granule cells: Protection by synthetic derivatives of endogenous sphingolipids, J. Pharmacol. Exp. Therap. **252:**419–427.

35. Higashi, H., Omozi, A., Yamagoto, T., 1992, Calmodulin, a ganglioside binding pr otein, J. Biol. Chem. **267:**9831–9838.

36. Watson, D. B., Dietrich, W. D., Busto, R., Wachtel, M. S., Ginstery, M. D., 1985, Induction of reproducible brain infarction by photochemically initiated thrombosis, Amer. Neurol. **17:**497–504.

37. Dietrich, W. D., Watson, B. D., Busto, R., Ginsberg, M. D., Bethea, J. R., 1987, Photochemically induced cerebral infartion I early microvascular alternations, Acta Neurol. **72:**315–334.

38. Costa, E., Kharlamov, A., Guidotti, A., Hayes, R., Armstrong, D., in press, Sequelae of biochemical events following photochemical injury of rat sensory–motor cortex: Mechanism of ganglioside protection, Physiopath. Exp. Therp.

CONTRIBUTORS

ARMSTRONG, D.
Fidia–Georgetown Institute
for the Neurosciences
Washington, D. C. 20007
U.S.A.

BASILE, A. S.
Liver Disease Section and
Laboratory of Neuroscience
NIDDK
National Institutes of Health
Bethesda, MD 20892
U.S.A.

CARDA, P.
Dto. Cirugía General y Digestiva
Hospital Ramón y Cajal
28034 Madrid
Spain

COOPER, A. J. L.
Dept. Biochemistry and Neurology
Cornell University
Medical College
New York, NY 10021
U.S.A.

COSTA, E.
Fidia–Georgetown Institute
for the Neurosciences
Washington, D. C. 20007
U.S.A.

FELIPO, V.
Instituto de Investigaciones
Citológicas de la F.I.B.
46010 Valencia
Spain

GRISOLIA, S.
Instituto de Investigaciones
Citológicas de la F.I.B.
46010 Valencia
Spain

GRAU, E.
Instituto de Investigaciones
Citologicas de la F.I.B
46010 Valencia
Spain

GUIDOTTI, A.
Fidia–Georgetown Institute
for the Neurosciences
Washington, D. C. 20007
U.S.A.

HAWKINS, R. A.
Dept. Physiology and Biophysics
University of Health Sciences
Chicago Medical School
North Chicago, IL 60064
U.S.A.

HERRERO, I.
Dept. Bioquímica
Facultad de Veterinaria
Universidad Complutense
28040 Madrid
Spain

JONES, E. A.
Liver Diseases Section and
Laboratory of Neurosciences
NIDDK
National Institutes of Health
Bethesda, Maryland 20892
U.S.A.

KHARMALOV, A.
Fidia–Georgetown Institute
for the Neurosciences
Washington, D. C. 20007
U.S.A.

KIEDROWSKI, L.
Fidia–Georgetown Institute
for the Neurosciences
Washington, D. C. 20007
U.S.A.

LAJTHA, A.
Nathan S. Kline Institute
for Psychiatric Research
Center for Neurochemistry
Orangeburg, NY 10962
U.S.A.

MANS, A. M.
Dept. Physiology and Biophysics
University of Health Sciences
Chicago Medical School
North Chicago, IL 60064
U.S.A.

MICHAELIS, E.
Dept. Pharmacology and Toxicology
Center for Neurobiology and
Immunology Research
University of Kansas
Lawrence, KS 66045
U.S.A.

MIRAS–PORTUGAL, M. T.
Dept. Bioquímica
Facultad de Veterinaria
Universidad Complutense
28040 Madrid
Spain

MIÑANA, M. D.
Instituto de Investigaciones
Citológicas de la F.I.B.
46010 Valencia
Spain

MOREJON, E.
Unidad de Nutrición Clínica y Dietética
Hospital Ramón y Cajal
28034 Madrid
Spain

RAABE, W.
Depts. Neurology and Physiology
V. A. Medical Center
University of Minnesota
Minneapolis, MN 55417
U.S.A.

RODES, J.
Hospital Clinic i Provincial
Unidad de Hepatología
Facultad de Medicina
Universidad de Barcelona
08036 Barcelona
Spain

SANCHEZ–PRIETO, J.
Dept. Bioquimica
Facultad de Veterinaria
Universidad Complutense
28040 Madrid
Spain

SASTRE, A.
Unidad de Nutricion Clinica y Dietética
Hospital Ramón y Cajal
28034 Madrid
Spain.

SKOLNICH, P.
Liver Diseases Section and
Laboratory of Neuroscience
NIDDK
National Institutes of Health
Bethesda, MD 20892
U. S. A.

WROBLEWSKI, J. T.
Fidia–Georgetown Institute
for the Neurosciences
Washington, D. C. 20007
U.S.A.

YURDAYDIN, C.
Liver Diseases Section and
Laboratory of Neuroscience
NIDDK
National Institutes of Health
Bethesda, MD 20892
U.S.A.

INDEX